Sungkyunkwan University Outstanding Research

Series Editor

Sukhan Lee
Dean of the Graduate School
SungKyunKwan University
Korea
E-mail: lsh@ece.skku.ac.kr

For further volumes:
http://www.springer.com/series/11431

Kwon-Ho Song and Cheorl-Ho Kim

Sialo-Xenoantigenic Glycobiology

Molecular Glycobiology
of Sialylglycan-Xenoantigenic
Determinants in Pig to Human
Xenotransplantation

Authors
Dr. Kwon-Ho Song
Molecular and Cellular Glycobiology
Laboratory
Department of Biological Science
SungKyunKwan University
Kyunggi-Do
Korea

Prof. Cheorl-Ho Kim
Molecular and Cellular Glycobiology
Laboratory
Department of Biological Science
Sungkyunkwan University
Kyunggi-Do
Korea

ISSN 2195-3546   e-ISSN 2195-3554
ISBN 978-3-642-43443-3   ISBN 978-3-642-34094-9 (eBook)
DOI 10.1007/978-3-642-34094-9
Springer Heidelberg New York Dordrecht London

Preface (Summary)

During the past two decades, glycobiology has hugely been developed in its biological significance and biological diversity. The glycobiology is specifically involved in cell-cell interaction, cell differentiation, receptor-mediated targeting, molecular recognition and transplantation. The scope of the present new book is narrow, focusing on carbohydrate antigens including sialic acids as xenoantigenic determinant in human.

In xenotransplantation, the pig has been identified as a suitable organ donor candidate for humans because of its compatible organ size and short breeding time (Scheme 1). However, exposure of pig organs to human blood results in hyper acute rejection (HAR) in pig to human xenotransplantation. The rejection is caused by differences in carbohydrate epitopes on the human and pig vascular endothelia (Table. 1). When pig organs or tissues are transplanted into the human body, the human IgM isotype of anti-Gal binds to Gal antigens on the pig tissues, which causes activation of the complement cascade resulting in cell lysis. The Gal antigen was eliminated by knocking out the α-1,3 galactosyltransferase, but the remaining so-called non-Gal antigens are considered to be xenoantigens subsequently involved in the rejection phenomenon.

Carbohydrate antigens, present on glycoconjugates of all mammalian cells, play crucial roles in various biological processes and are epitopes recognized by the immune system. Among them, carbohydrate antigens containing sialic acid, such as sialosyl-Tn or Hanganutziu-Deicher (HD), are non-Gal antigens against which humans are suggested to have naturally occurring antibodies (Table. 2).

To overcome rejection responses such as HAR in xenotransplantation, studies of genes involved in carbohydrate antigens that cause xenoantigenicity are necessary. Knowledge of pig glycosyltransferases would be useful to apply to xenoantigen masking or identification of the xenoantigenic sialylglycan(s). However, most pig glycosyltransferase genes have not yet been isolated. Therefore, in the first chaprer of the present study, we screened for pig glycosyltransferase genes involved in generating xenoantigens. In the chpter II to IV, we cloned, functionally characterized, and investigated the regulatory mechanism of the pig CMAH gene in NeuGc biosynthesis. Lastly, we investigated the effects of an alteration of pig glycosylation pattern on human serum-mediated cytotoxicity, caused by human sialyltransferases including hST6GalNAc IV.

Keywords: *glycosyltransferase, sialyltransferase, xenoantigen, N-glycolylneuraminic acid*

Xenotransplantation and the Rejection Spectrum

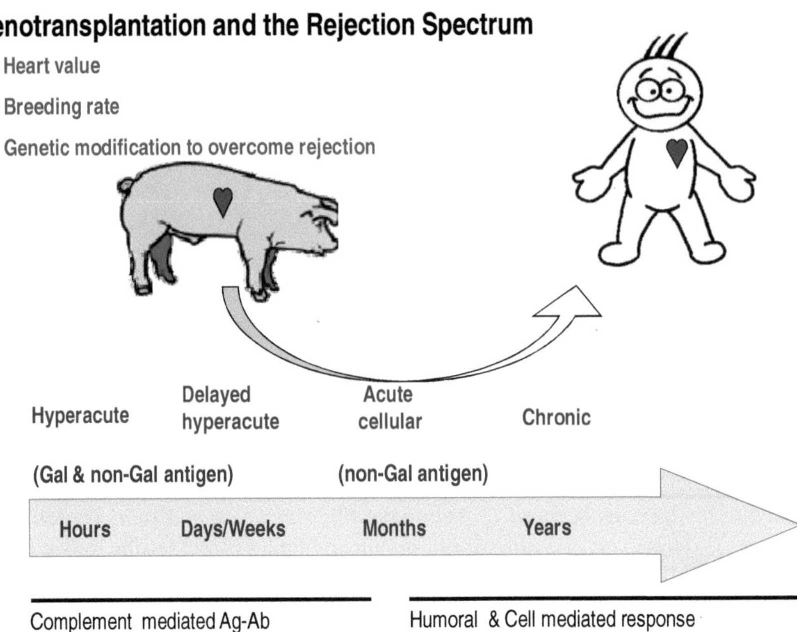

Scheme 1. Xenotransplantation and the rejection spectrum

Table 1. Structure of the known carbohydrate epitopes exposed at the surface of human and pig vascular endothelia

Human	Pig
Galβ 1,4GlcNAcβ 1-R	Galβ 1,4GlcNAcβ 1-R
ABH-Galβ 1,4GlcNAcβ 1-R[a]	**Galα 1,3**Galβ 1,4GlcNAcβ 1-R[b]
NeuAcα 2,3Galβ 1,4GlcNAcβ 1-R[c]	NeuAcα 2,3Galβ 1,4GlcNAcβ 1-R[c]
	NeuGcα 2,3Galβ 1,4GlcNAcβ 1-R[d]

Modified from Oriol et al.
The epitopes represented by bold type are different between two species.
R is glycolipid or glycoprotein carrier molecules anchored in the cell membrane.
[a] A,B,H or AB blood group antigen. [b] Gal antigen. [c] N-acetylneuramic acid.
[d] N-glycolylneuramic acid.

Table 2. Known carbohydrate antigens against which humans may have naturally occurring antibodies

1.   **Hanganutziu-Deicher : NeuGc**

2.   **Thomsen-Friedenreich (T or TF) : Galβ1,3GalNAcα1-R**

3.   **Tn(TF precursor) : GalNAcα1-R**

4.   **Sialosyl-Tn : NeuAcα2,6GalNAcα1-R**

5.   **Forssman : GalNAcα1,3GalNAcβ1,3Galα1,4Galβ1,4Glcβ1-R**

6.   **αRhamnose-containing oligosaccharides**

7.   **Sulphatide I : SO_4-3Gal-R**

Modified from Ezzelarab et al. R is glycolipid or glycoprotein carrier molecules anchored in the cell membrane

Contents

List of Figures

Chapter 1

Chapter 2

Chapter 3

Chapter 4

Chapter 5

List of Tables

Chapter 1
Screening of Pig Glycosyltransferase Genes Related to Xenoantigens and Their Masking

Kwon-Ho Song and Cheorl-Ho Kim[*]

Molecular and Cellular Glycobiology Laboratory, Department of Biological Science, SungKyunKwan University, 300 Chunchun-Dong, Jangan-Gu, Suwon City, Kyunggi-Do 440-746, Korea
kwonho@live.co.kr, chkimbio@skku.edu

Abstract. In this study, to screen for pig glycosyltransferase genes involved in xenoantigen synthesis or masking, pig EST sequences were collected from the TIGR and NCBI databases, and local BLAST was performed using a known human sialyltransferase gene. Sequence fragments of pig sialyltransferase genes were obtained from the BLAST results, and from these, full ORF sequences of three pig sialyltransferase genes, ST3Gal III, ST3Gal IV, and ST8Sia IV, were isolated. A partial fragment of the pig iGb3 synthase gene was also isolated by PCR, using primers based on the conserved sequence information of known iGb3S genes.

1 Introduction

Glycoconjugates are generated by the sequential and coordinated action of many glycosyltransferases (Milland, et al., 2005). Glycosyltransferases, located in the endoplasmic reticulum (ER) and Golgi apparatus, catalyze the sequential transfer of monosaccharides from nucleotide sugars to saccharide acceptors, resulting in mature oligosaccharides (Milland et al., 2005). Many glycosyltransferases are involved in xenotransplantation, related to the expression or masking of xenoantigens. Among these, the most important glycosyltransferases in xenotransplantation are the α1,3 galactosyltransferases including the α1,3GT and iGb3 synthase (iGb3S).

α1,3GT and iGb3S both transfer UDP-Gal but differ in the acceptor utilized: N-acetyllactosamine (NacLac, protein and lipid) or lactosylceramide (LacCer, lipid), respectively (Fig. 1). A terminal galactose-α1,3-galactose (Gal) residue is acommon structure of carbohydrate antigens presented on most mammalian cell surfaces, well characterized because of their importance in blood transfusion and

---

[*] Corresponding author.

K.-H. Song and C.-H. Kim: *Sialo-Xenoantigenic Glycobiology*, SKKU 1, pp. 1–10.
DOI: 10.1007/978-3-642-34094-9_1   © Springer-Verlag Berlin Heidelberg 2013

organ allo- and xeno-transplantation (Heissigerova et al., 2003). The Gal-α1,3-Gal antigen (Gal antigen) is a major xenoantigen in pig-to-human xenotransplantation. The pig α1,3GT gene, was isolated and targeted to develop α1,3GT deficient pigs. However, the Gal-α1,3-Gal antigen is still present in α1,3GT deficient pigs and possibly synthesized on lipids by iGb3S. Therefore, removal of the pig iGb3S gene is the next step necessary to achieve successful pig-to-human xenotransplantation without rejection responses. Although iGb3S has been cloned and characterized from mouse and rat (Taylor et al., 2003; Milland et al., 2006), the pig iGb3S gene has not yet been cloned.

Sialyltransferases are a family of more than 18 members that catalyze the transfer of sialic acid from CMP-Neu5Ac forming an α2,3-, an α2,6-, or an α2,8-linkage, depending on the acceptor sugar chain. In pig-to-human xenotransplantation, sialyltransferases such as ST3Gal III and ST6Gal I reduce the levels of Gal antigen by competing with α1,3-galactosyltransferase for the common acceptor substrate (Ezzelarab and Cooper, 2005; Koma et al., 2000; Tanemura et al., 1998). Almost all members of the sialyltransferase gene family of human and mouse have been isolated and characterized. However, most pig sialyltransferase genes have not yet been isolated, even though pig sialyltransferases would be useful to apply to xenoantigen masking or identification of xenoantigenic sialylglycan(s).

In this study, to screen for pig glycosyltransferase genes involved in xenoantigen synthesis or masking, pig EST sequences were collected from the TIGR and NCBI databases, and local BLAST was performed using a known human sialyltransferase gene. Sequence fragments of pig sialyltransferase genes were obtained from the BLAST results, and from these, full ORF sequences of three pig sialyltransferase genes, ST3Gal III, ST3Gal IV, and ST8Sia IV, were isolated. A partial fragment of the pig iGb3 synthase gene was also isolated by PCR, using primers based on the conserved sequence information of known iGb3S genes.

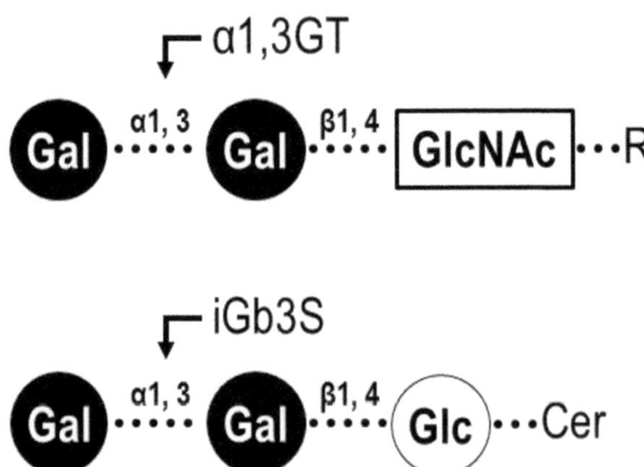

Fig. 1. Two types of α1, 3 galactosyltransferase involved in biosynthesis of Gal antigen R is glycolipid or glycoprotein carrier molecules anchored in the cell membrane. Gal; Galactose, Glc; Glucose, GlcNAc; N-acetylglucosamin, Cer; Ceramide.

2   Materials and Methods

2.1   Local BLAST

Pig expressed sequence tag (EST) sequences were collected from the institute of genome research (TIGR, http://www.tigr.org) and national center for biotechnology information (NCBI, www.ncbi.nlm.nih.gov) databases. Local BLAST searches were performed using BioEdit software to obtain the sequence information of pig sialyltransferase genes corresponding to the following human sialyltransferase genes (accession number in parentheses) : ST6Gal I (X17247), ST6GalNAc I (Y11339), ST6GalNAc II (AJ251053), ST6GalNAc IV (AJ271734), ST3Gal I (L29555), ST3Gal II (U63090), ST3Gal III (L23768), ST3Gal IV (L23767), ST3Gal V (AB018356), ST3Gal VI (AB022918), ST8Sia I (D26360), ST8Sia II (U33551), ST8Sia III (AF004668), ST8Sia IV (L41680), and ST8Sia V (U91641).

2.2   Cell Culture

A pig kidney cell line (PK15) was obtained from the Korean Cell Line Bank (KCLB) and pig ear fibroblast (PEF) was kindly provided by the National Livestock Research Institute, RDA. Cells were grown in Dulbecco's modified Eagle's medium (DMEM; WelGENE, Korea) containing 100 units/ml of penicillin-streptomycin and 10% fetal bovine serum (FBS; WelGENE, Korea) at 37°C in a 5% CO_2 incubator/humidified chamber.

2.3   Isolation of Pig Sialyltransferases and iGb3S Gene

Total RNA was isolated from PK15 cells, PEF cells, or fresh pig tissues using TRIZOL reagent (Invitrogen, Carlsbad, CA), and cDNAs were synthesized by reverse transcriptase (RT) with an oligo dT-adaptor primer using AccuPower® RT-PreMix (Bioneer Co., Korea). For cloning the pig sialyltransferase, PCR was performed on PK15 orPEF cDNA using the following primer sets: ST3Gal I, 5'-ACCATGGCCCCCATGAGGAAGAAG A-3' (sense) and 5'-ACCTCATCTGCCCTTGAAGATCCGGA-3' (antisense); ST3Gal III, 5'-ACCATGGGACTCTTGGTATTTGTACG-3' (sense) and 5'-ACCTCAGATGCCGCTGGTCAGGTCA-3' (antisense) ST3Gal IV, 5'-ACCATGATCAGCAAGTGCCGCTGGA-3' (sense) and 5'-ACCTCAGAAGTATGTGAGGTTCTTG-3' (antisense); ST6GalNAc IV, 5'-ACCATGAAGCCTCCGGGTCGGCTCC-3' (sense) and 5'-ACCCTACTGGGTCTTCCAGGAGGGG-3' (antisense); ST8Sia IV, 5'-ACCATGCGCTCCATTAGGAAGAGGTG-3' (sense) and 5'-ACCTTATTGCTTTACGCACTTTCCTG-3' (antisense). The pig iGb3 gene fragment was amplified by PCR using the following primer set: iGb3S-S,

5'-GTGTTCTGCCTGGACGTG-3'   and   iGb3S-AS,   5'-CCAGAAGAACTTGTTAAGGTGG CT-3'. The amplified DNA fragments were subcloned into the EcoRV site of pSTBlue-1 AccepTorTM vector (Novagen). Multiple independent clones were isolated and sequenced to confirm the sequence data.

2.4 Reverse Transcription Polymerase Chain Reaction (RT-PCR)

Total RNA was isolated from various pig tissues and cDNA was synthesized with oligo dT primer using 1 ug of total RNA as template. cDNA from various pig tissues was amplified by PCR with the above primer sets using EF-Taq polymerase (SolGent, Korea). The use of equal amounts of mRNA in the RT-PCR assay was confirmed by assessing β-actin expression levels.

Pig EST sequences were collected from the TIGR and NCBI data base, and Local blast was performed using a known human sialyltransferase gene. Black bar indicates sequence data from TIGR and gray bar indicates sequence data from NCBI. Overlapped region between TIGR and NCBI sequence data was represented by black bar. The number indicates sequence length.

3 Results

3.1 Screening and Cloning of Pig Sialyltransferase Genes

The nucleotide sequences of 17 human sialyltransferase genes were used to perform a local BLAST analysis of pig EST sequences in the TIGR and NCBI data base. As a result, pig EST sequences corresponding to each query sequence were collected. The retrieved sequences were aligned with the query sequence, and partially overlapping sequences were assembled to build a larger fragment or ORF (Table 1). From this analysis, we obtained full ORF sequences of five pig sialyltransferase genes, ST3Gal I, ST3Gal III, ST3Gal IV, ST6GalNAC IV, and ST8Sia IV, and sequence fragments for 9 pig sialyltransferases. We did not find pig ESTs for ST6GalNAc I, ST6GalNAc III, and ST8Sia VI.

Next, to isolate the complete gene sequences of the five pig sialyltransferase genes, ST3Gal I, ST3Gal III, ST3Gal IV, ST6GalNAC IV, and ST8Sia IV, we performed PCR using PK15 or PEF cDNA as template. As a result, the pig ST3Gal III and ST3Gal IV gene were isolated from PK15 cells, and the ST8Sia IV gene was isolated from PEF cells (Fig. 2). The pig ST3Gal III gene (Fig. 3) is comprised of 1077 bp encoding 358 amino acids, pig ST3Gal IV (Fig. 4) has 1002 bp encoding 333 amino acids, and pig ST8Sia IV (Fig. 5) has 1080 bp encoding 359 amino acids. Multiple alignment analysis revealed that three sialylmotifs were highly conserved in these 3 sialyltransferase genes (Fig. 6).

Table 1. Local BLAST using Human sialyltransferase gene

Gene	sequence	Alignment positions (scale 0.1–1.2 kb)
ST6Gal I	1121	1 – 492
ST3Gal I	1023 (1032)	49, 147, 167 … 1023
ST3Gal II	1053	608 … 1053
ST3Gal III	1128 (1125)	1 … 1128
ST3Gal IV	999 (1002)	10 … 999
ST3Gal V	1089	1, 211, 292 … 905, 967, 1073
ST3Gal VI	996	1 … 588
ST6GalNAc I		
ST6GalNAc II	1125	1, 106, 217 … 401, 445 … 1125
ST6GalNAc III		
ST6GalNAc IV	909 (909)	1 … 898
ST8Sia I	1026	111 … 1026
ST8Sia II	1128	109 … 550
ST8Sia III	1143	179 … 300, 837 … 1143
ST8Sia IV	1080 (1089)	1 … 713
ST8Sia V	1131	1 … 197
ST8Sia VI		

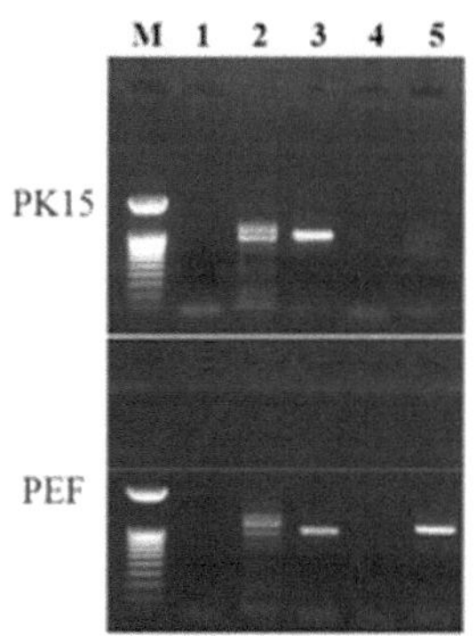

Fig. 2. Isolation of pig sialyltransferase by PCR from the pig cell lines, PK15 and PEF.
M: 100 bp ladder marker, 1: ST3Gal I (1032 bp), 2: ST3Gal III (1077 bp), 3: ST3Gal IV
(1002 bp), 4: ST6GalNAc IV (909 bp), 5: ST8Sia IV (1080 bp).

```
atgggactcttggtatttgtacgcaatctgctgctagccctctgcctttttctggtgttg
 M  G  L  L  V  F  V  R  N  L  L  L  A  L  C  L  F  L  V  L
ggattttgtattattctgcgtggaagctacatttactccagtgggaggactccaagtat
 G  F  L  Y  Y  S  A  W  K  L  H  L  L  Q  W  E  D  S  K  Y
gaccggttgggctttctcctgaagctggactctaaactgcctgctgagttggccaccaag
 D  R  L  G  F  L  L  K  L  D  S  K  L  P  A  E  L  A  T  K
tacgcaaacttttcagaaggagcttgcaagcctggctatgcgtcagccgtgatgactgcc
 Y  A  N  F  S  E  G  A  C  K  P  G  Y  A  S  A  V  M  T  A
atcttccccggttctccaaaccagcacccatgttcctggatgactccttccgaaagtgg
 I  F  P  R  F  S  K  P  A  P  M  F  L  D  D  S  F  R  K  W
gccaggatccgggaatttgtgccgccttttgggatcaaaggtcaagacaatctgatcaga
 A  R  I  R  E  F  V  P  P  F  G  I  K  G  Q  D  N  L  I  R
gccatcctgtccgtcaccaaagagtaccgcctgaccccggccttggacagcctcagctgc
 A  I  L  S  V  T  K  E  Y  R  L  T  P  A  L  D  S  L  S  C
cgccgctgcatcatcgtgggcaacggaggtgtcctcgccaacaagtctctggggtcacga
 R  R  C  I  I  V  G  N  G  G  V  L  A  N  K  S  L  G  S  R
attgatgactatgacattgtggtcagactgaactcggcgccagtgaaaggcttcgagaag
 I  D  D  Y  D  I  V  V  R  L  N  S  A  P  V  K  G  F  E  K
gacgtgggcagcaaaaccacactacgcatcacctaccccgagggcgccatgcagcggccc
 D  V  G  S  K  T  T  L  R  I  T  Y  P  E  G  A  M  Q  R  P
gagcaatacgaacgcgattctctcttcgtcctcgctggcttcaagtggcaggacttcaag
 E  Q  Y  E  R  D  S  L  F  V  L  A  G  F  K  W  Q  D  F  K
tggctgaagtacatcgtctacaaggagcgagtgagtgcatcggacggcttctggaaatcg
 W  L  K  Y  I  V  Y  K  E  R  V  S  A  S  D  G  F  W  K  S
gtggccactcgcgtgcccaaggaaccccagagattcgcatcctcaatccatatttcatc
 V  A  T  R  V  P  K  E  P  P  E  I  R  I  L  N  P  Y  F  I
caggaagccgccttcaccctcatcggactgcctttcaacaatggcctcatgggccgaggg
 Q  E  A  A  F  T  L  I  G  L  P  F  N  N  G  L  M  G  R  G
aacatcccgaccctcggcagtgtggccgtgaccatggcactgcacggctgcgatgaggtg
 N  I  P  T  L  G  S  V  A  V  T  M  A  L  H  G  C  D  E  V
gccgtcgccggcttcggctacgacatgagcacacccaatgcaccctgcactactacgag
 A  V  A  G  F  G  Y  D  M  S  T  P  N  A  P  L  H  Y  Y  E
accgtgcgcatggcagccatcaaagagtcctggacgcacaacatccagcgagagaaagag
 T  V  R  M  A  A  I  K  E  S  W  T  H  N  I  Q  R  E  K  E
tttctgcagaagctggtcaaagcacgcgtcatcactgacctgaccagcggcatctga
 F  L  Q  K  L  V  K  A  R  V  I  T  D  L  T  S  G  I  -
```

Fig. 3. Nucleotide sequence and deduced amino acid sequence of the pig ST3Gal III. The pig ST3Gal III gene was consisted of 1077 bp encoding 358 amino acids. The deduced amino acid sequence was shown below the nucleotide sequence.

```
atgatcagcaagtgccgctggaagctcctagccatgttggctttcgtcctggtggtcatggtatggtactccatc
 M  I  S  K  C  R  W  K  L  L  A  M  L  A  F  V  L  V  V  M  V  W  Y  S  I
tcccgagaagacagatacattgagctttttttattttcccatcctagagaagaaggagccatgcctccaggggggag
 S  R  E  D  R  Y  I  E  L  F  Y  F  P  I  L  E  K  K  E  P  C  L  Q  G  E
gcggagaggaaggtctccaagctctttggcaactactcccaagaacagcccatcttcctgcaacttaaggattac
 A  E  R  K  V  S  K  L  F  G  N  Y  S  Q  E  Q  P  I  F  L  Q  L  K  D  Y
ttctgggtcaagacgccatctgcttacgagctgccctacgggaccaaggggagtgaagacctgcttctccgggtg
 F  W  V  K  T  P  S  A  Y  E  L  P  Y  G  T  K  G  S  E  D  L  L  L  R  V
ctggccatcaccagctattccattccagagagcatccagaggctcaagtgccgccgctgcgtggtggtggggaat
 L  A  I  T  S  Y  S  I  P  E  S  I  Q  R  L  K  C  R  R  C  V  V  V  G  N
gggcatcggctgcgcaacagcttgctgggagaggccatcaacaagtacgacgtggtcatcaggctaaacaatgcc
 G  H  R  L  R  N  S  L  L  G  E  A  I  N  K  Y  D  V  V  I  R  L  N  N  A
ccggtggctggctacgagggtgacgtgggctctaagaccaccatgcgtctcttctaccctgaatctgcccacttc
 P  V  A  G  Y  E  G  D  V  G  S  K  T  T  M  R  L  F  Y  P  E  S  A  H  F
aacoccaaagtggaggacaacccggacacactcctcgtcctggtagctttcaaggccatggacttccactggatt
 N  P  K  V  E  D  N  P  D  T  L  L  V  L  V  A  F  K  A  M  D  F  H  W  I
gagagcatcttgagtgacaagaagcgggttcgaaagggtttctggaaacagcctcccctcatctgggacgtcaac
 E  S  I  L  S  D  K  K  R  V  R  K  G  F  W  K  Q  P  P  L  I  W  D  V  N
cccaaacagattcgaattctcaacccccttcttcatggagattgcagctgacaaactgctaaggctgccgatgcaa
 P  K  Q  I  R  I  L  N  P  F  F  M  E  I  A  A  D  K  L  L  R  L  P  M  Q
cagccacacaagattaagcagaagcccaccacgggcctgctggccatcaccctggccctccacctgtgtgacctg
 Q  P  H  K  I  K  Q  K  P  T  T  G  L  L  A  I  T  L  A  L  H  L  C  D  L
gtgcacattgctggcttcggctacccagatgcccaaaacaagaaggagtccattcactactatgagcaaatcacc
 V  H  I  A  G  F  G  Y  P  D  A  Q  N  K  K  E  S  I  H  Y  Y  E  Q  I  T
atcaagtccatggcgtggtcaggccacaatgtctcccaagaggccctggccatcaagcggatgctggaaatcgga
 I  K  S  M  A  W  S  G  H  N  V  S  Q  E  A  L  A  I  K  R  M  L  E  I  G
gcggtcaagaacctcacatacttctga
 A  V  K  N  L  T  Y  F  -
```

Fig. 4. Nucleotide sequence and deduced amino acid sequence of the pig ST3Gal IV. The pig ST3Gal IV gene was consisted of 1002 bp encoding 333 amino acids. The deduced amino acid sequence was shown below the nucleotide sequence.

```
aggaagaggtggacgatctgcacgataagtctgctcctgatcttttataagacaaaa
 R  K  R  W  T  I  C  T  I  S  L  L  L  I  F  Y  K  T  K
actgaggagcaccaggagacgcaactcatcggagatggcgagctgtgtttgagtcga
 T  E  E  H  Q  E  T  Q  L  I  G  D  G  E  L  C  L  S  R
agttcagataaaatcattcgaaaggctggctcttcaatcttccagcattctgtagaa
 S  S  D  K  I  I  R  K  A  G  S  S  I  F  Q  H  S  V  E
aattcctctttggtcctggagataaggaagaacattcttcgtttccttgatgcagaa
 N  S  S  L  V  L  E  I  R  K  N  I  L  R  F  L  D  A  E
gtggtcaagagcagttttaagcctggtgatgtcatacactatgttctggataggcgc
 V  V  K  S  S  F  K  P  G  D  V  I  H  Y  V  L  D  R  R
atttctcaggatctgcacagcctcctacctgaagtttcaccaatgaaaaatcgcaga
 I  S  Q  D  L  H  S  L  L  P  E  V  S  P  M  K  N  R  R
gcagttgttgggaattctggcattctgctagatagtgaatgtggaaaagagattgac
 A  V  V  G  N  S  G  I  L  L  D  S  E  C  G  K  E  I  D
gtaataaggtgtaatctagcccctgtggtggagtttgctgcagatgtgggaactaaa
 V  I  R  C  N  L  A  P  V  V  E  F  A  A  D  V  G  T  K
accatgaatccatcagttgtacaaagagcatttggaggctttcggaatgaaagtgac
 T  M  N  P  S  V  V  Q  R  A  F  G  G  F  R  N  E  S  D
gtgcacagactttccatgctgaatgacagtgtcctttggatccctgctttcatggtc
 V  H  R  L  S  M  L  N  D  S  V  L  W  I  P  A  F  M  V
aagcacgtggagtgggttaatgcattaatccttaagaataaactgaaagtgcgaact
 K  H  V  E  W  V  N  A  L  I  L  K  N  K  L  K  V  R  T
ctgagacttattcatgctgtcagaggttactggctgacaaacaaagttcctatcaaa
 L  R  L  I  H  A  V  R  G  Y  W  L  T  N  K  V  P  I  K
ggtctgctcatgtatacacttgccacgagattctgtgatgaaattcacctgtatgga
 G  L  L  M  Y  T  L  A  T  R  F  C  D  E  I  H  L  Y  G
cctaaggatttgaatggaaaagctgtcaaatatcattactatgatgacttaaaatac
 P  K  D  L  N  G  K  A  V  K  Y  H  Y  Y  D  D  L  K  Y
aatgcaagtcctcatagaatgccattagaattcaaaacattaaatgtgctgcataat
 N  A  S  P  H  R  M  P  L  E  F  K  T  L  N  V  L  H  N
aaactgacaacaggaaagtgcgtaaagcaataa
 K  L  T  T  G  K  C  V  K  Q  -
```

Fig. 5. Nucleotide sequence and deduced amino acid sequence of the pig ST8Sia IV. The pig ST8Sia IV gene was consisted of 1080 bp encoding 359 amino acids. The deduced amino acid sequence was shown below the nucleotide sequence.

```
ST3Gal III   MGLLVFVRNLLLALCLFLVLGFLYYSAWKLHLLQWEDSKYDRLGFLLKLDSKLPAELATK  60
ST3Gal IV    --MISKCRWKLLAMLAFVLVVMVWYS------ISRED-RYIELFYFPILEKKEP------  45
ST8Sia IV    -MRSIRKRWTICTISLLLIFYKTKEIARTEEHQETQLIGDGELCLSRSLVNSSDKIIRKA  59

ST3Gal III   YANFSEGACKPGYASAVMTAIFPRFSKPAPMFLDDSFRKWARIR-EFVPPFGIKGQDNLI  119
ST3Gal IV    --------CLQGEAERKVSKLFGNYSQEQPIFLQLKDYFWVKTPSAYELPYGTKGSEDLL  97
ST8Sia IV    G----SSIFQHSVEGWKINSSLVLEIRKNILRFLDAERDVSVVKSSFKPGDVIHYVLDRR  115
                                                             Large motif
ST3Gal III   RAILSVTKEYRLTPALDSLSCR---RCIIVGNGGVLANKSLGSRIDDYDIVVRLNSAPVK  176
ST3Gal IV    LRVLAIT-SYSIPESIQRLKCR---RCVVVGNGHRLRNSLLGEAINKYDVVIRLNNAPVA  153
ST8Sia IV    RTLNISQDLHSLLPEVSPMKNRRFKTCAVVGNSGILLDSECGKEIDSHNFVIRCNLAPVV  175

ST3Gal III   GFEKDVGSKTTLRITYPEGAMQRP---EQYER-DSLFVLAGFKWQDFKWLKYIVYKERVS  232
ST3Gal IV    GYEGDVGSKTTMRLFYPESAHFNP---KVEDNPDTLLVLVAFKAMDFHWIESILSDKKR-  209
ST8Sia IV    EFAADVGTKSDFITMNPSVVQRAFGGFRNESDREKFVHRLSMLNDSVLWIPAFMVKGGE-  234

ST3Gal III   ASDGFWKSVATRVPKEPPEIRILNPYFIQEAAFTLIGLPFNNGLMGRGNIPTLGSVAVTM  292
ST3Gal IV    VRKGFWKQPPLIWDVNPKQIRILNPFFMEIAADKLLRLPMQQPHKIK-QKPTTGLLAITL  268
ST8Sia IV    -KHVEWVNALILKN--KLKVRTAYPSLRLIHAVRGYWLTNKVPIK----RPSTGLLMYTL  287
               Small motif                            Very small motif
ST3Gal III   ALHGCDEVAVAGFGY-----DMSTPNAPLHYYETVRMAAIKESWTHNIQR--------EK  339
ST3Gal IV    ALHLCDLVHIAGFGYP----DAQNKKESIHYYEQITIKSMAWSG-HNVSQ--------EA  315
ST8Sia IV    ATRFCDEIHLYGFWPFP--KDLNGKAVKYHYYDDLKYRYFSNASPHRMPL--------EF  337

ST3Gal III   EFLQKLVKARVITDLTSGI---  358
ST3Gal IV    LAIKRMLEIGAVKNLTYF----  333
ST8Sia IV    KTLNVLHNRGALKLTTGKCVKQ  359
```

Fig. 6. Allignment of pig sialyltransferase protein sequences and conserved sialylmotifs. Multiple allignments of the deduced amino acid sequence of three pig sialyltransferases were analysed by using the ClustalW program. Conserved sialylmotifs (Large, Small and Very small) were underlined.

3.2 *Isolation of the Partial Pig iGb3S Gene*

To isolate the pig iGb3S gene, iGb3S gene sequences from other animals were used to generate primers for PCR. A partial 439 bp fragment of the iGb3S gene was isolated by PCR from pig tissue-mixture RNA and sequenced (Fig. 7). A multiple alignment analysis of the deduced pig iGb3S amino acid sequence with that of various species such as cattle, marmoset, dog, rat, mouse and human was performed (Fig. 8). As shown in Table 2, the deduced amino acid sequence of pig iGb3S has high identities with that of various species. The pig iGb3S amino acid sequence exhibited 85% identity with cattle, 86% with marmoset, 85% with dog, 79% with mouse and rat, and 82% with human (Table 2). These results indicate that the partial pig iGb3S gene has high similarities with various iGb3S genes.

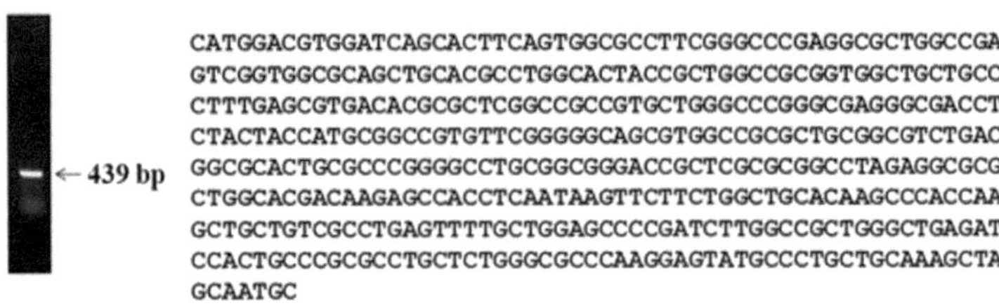

Fig. 7. Isolation of partial iGb3 synthase (iGb3S) gene from PK15 genomic DNA.
(A) Pig partial iGb3S gene was isolated by PCR from PK15 genomic DNA. (B) Nucleotide sequences of pig partial iGB3S gene.

```
Macaca mulatta       241 GRLGREAHFVFCMDVDQHFSSTFGPEALAESVAQLHSWHYHWPRWLLPFERDAHSAAAMARGEGDFYYHAAVFGGSVAAL 320
Callithrix jacchus   222 GWLGREAHFVFCMDIDQHFSGAFGPEALAESVAQLHSWHYHWPRWLLPFERDARSAAALTRDEGDFYYHAAVFGGSVAAL 301
Canis familiaris     135 GRLGREAHFVFCMDVDQHFRSSFGPEALAESVAQLHAWHYHRPRWLLPYERDARSAAAVAPGEGDFYYHAAVFGGSVAAL 214
Homo sapiens         178 GLPGREAHFMFCMDVDQHFSGTFGPEALAESVAQLHSWHYHWPSWLLPFERDAHSAAAMAWGQGDFYNHAAVFGGSVAAL 257
Rattus norvegicus    185 GQLGREADYVFCLDVDQYFSGNFGPEVLADLVAQLHAWHFRWPRWMLPYERDKRSAAALSLSEGDFYYHAAVFGGSVAAL 264
Mus musculus         216 GQLGQEADFVFCLDVDQYFTGNFGPEVLADLVAQLHAWHYRWPRWLLPYERDKRSAAALSLSEGDFYYHAAVFGGSVAAL 295
Sus scrofa             1 ---------VFCMDVDQHFSGAFGPEALAESVAQLHAWHYRWPRWLLPFERDTRSAAVLGPGEGDLYYHAAVFGGSVAAL 71
                         :**:*:**:*  . ****.**: *****:**:: * *:**:*** :***.:   .:**:* ***********

Macaca mulatta       321 RDLTAHCARGLAWDRARGLEARWHDESHLNKFFWLHKPAKVLSPEFCWSPDLGRRAEIRRPRLLWAPKEYRLLRD 395
Callithrix jacchus   302 RGLTAHCALGLAWDRARGLEARWHDESHLNKFFWLHKPAKVLSPEFCWSPDIGPRAEIRRPRLLWAPKEYRLLRN 376
Canis familiaris     215 RGLTGHCARALQRDRERGLEARWHDESHLNKFFWLHKPAKVLSPEFCWSPDIGWRAEIRRPRLLWAPKEYALVRD 289
Homo sapiens         258 RGLTAHCAGGLDWDRARGLEARWHDESHLNKFFWLHKPAKVLSPEFCWSPDIGPRAEIRRPRLLWAPKGYRLLRN 332
Rattus norvegicus    265 LKLTAHCATGQQLDREHGIEARWHDESHLNKFFWLSKPTKLLSPEFCWAEEIGWRPEIHHPRLIWAPKEYALVRT 339
Mus musculus         296 LKLTAHCATGQQLDHKRGIEALWHDESHLNKFFWLNKPTKLLSPEFCWAEEIIWRREIHHPRLLWAPKEYTLVRN 370
Sus scrofa            72 RRLTAHCARGLRRDRSRGLEARWHDKSHLNKFFWLHKPTKLLSPEFCW-------------------------- 119
                         **.***  .   *: :*:** ***:********* **:*:*******
```

Fig. 8. Multiple alignment of iGb3S deduced amino acid sequence of pig with that of other species.
Multiple alignment was analyzed by using the Clustal W program. (*) Fully conserved residues, (:) conservation of strong groups and (.) conservation of weak groups. Amino acid numbering is indicated on end of the right and left.

Table 2. Identities of pig iGb3S deduced amino acid sequence with that of other species

Organism	Identities (%)	Accession no.
Macaca mulatta	85	XP_001099773
Callithrix jacchus	86	XP_002750717
Canis lupus familiaris	85	XP_544424
Homo sapiens	82	CAI13954
Rattus norvegicus	79	NP_612533
Mus musculus	79	NP_001009819, XP_144044

3.3   Expression of Pig Glycosyltransferase Genes in Pig Tissues

To investigate tissue specific expression patterns of pig glycosyltransferase genes such as ST3Gal III, ST3Gal IV, ST8Sia IV, or iGb3S, total RNA was isolated from various pig tissues and the resulting cDNAs were prepared for RT-PCR analysis. The pig sialyltransferase and iGb3S genes were found to exhibit distinct tissue specific expression patterns (Fig.9.). The pig ST3Gal III gene was highly expressed in brain, muscle, testicle, kidney and heart, whereas the pig ST3Gal IV gene mainly expressed in spleen and testicle. In contrast, the pig ST8Sia IV gene has low expression levels, with weak expression in brain, rectum, tongue, muscle, spleen, small intestine, testicle, kidney, and heart. Because the pig iGb3S gene was expressed at a level too low for detectionin all pig tissues, we performed PCR using additional amplification cycles (42 cycles). As a result, we found that the pig iGb3S gene is highly expressed in brain, testicle, liver and kidney.

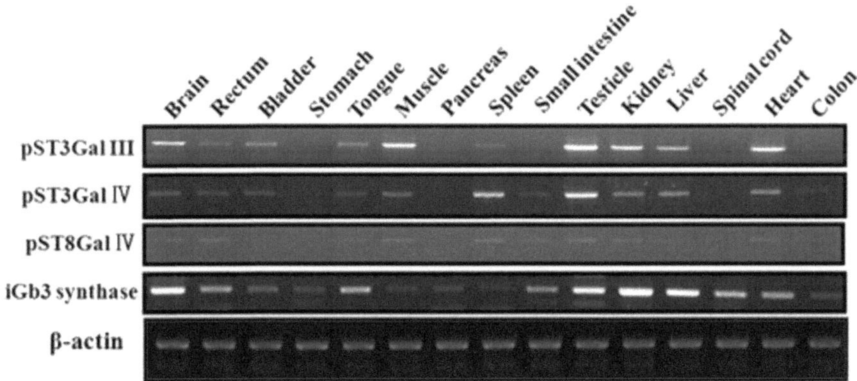

Fig. 9. mRNA expression of pig sialyltransferase genes and iGb3 synthase gene on various pig tissues.

Relative expression levels of pig sialyltransferase genes and iGb3 synthase gene were measured by RT-PCR in various pig tissues. β-Actin was used as a control.

4 Discussion

Carbohydrate antigens including the Gal antigen are important in pig-to-human xenotransplantation and genetic approaches to modify these carbohydrate antigens have also progressed. To overcome the rejection responses in pig-to-human xenotransplantation, the study of glycosyltransferase biology and modification of cell surface carbohydrates is necessary. However, the nucleotide sequence information of the pig glycosyltransferase genes involved in carbohydrate antigen biosynthesis of carbohydrate antigens was little studied.

Here, we obtained sequence fragments of 14 pig sialyltransferase genes using local BLAST analysis of the TIGR and NCBI pig ETS database. Among these, the sequences of five genes contained the complete ORF sequence, and cDNAs corresponding to the ORF sequences of ST3Gal III, ST3Gal IV, and ST8Sia IV were isolated from the pig cell lines, PK15 and PEF. ST3Gal III mediates the transfer of sialic acid residues to a Gal residue of terminal Galβ1-3GlcNAc oligosaccharide and is therefore the candidate for the synthesis of sialyl-LewisA epitope. ST3Gal IV mediates the transfer of sialic acid residues to a Gal residue of glycolipids or glycoproteins containing Galβ1-4GlcNAc or Galβ1-3GalNAc sequences. ST8Sia IV shows polysialic acid synthase activity *in vitro* towards N-CAM. These three sialyltransferase genes have three highly conserved sialylmotifs. In addition, the iGb3S gene was partially isolated from pig tissues. Three sialyltransferase genes and the iGb3S gene show tissue specific expression patterns. The sialyltransferase and iGb3S gene sequences acquired from this study will be useful in studying xenoantigens and xenoantigen masking.

Chapter 2
Cloning and Tissue Specific Expression of *pcmah* and Its Alternative Transcripts

Kwon-Ho Song and Cheorl-Ho Kim[*]

Molecular and Cellular Glycobiology Laboratory, Department of Biological Science,
SungKyunKwan University, 300 Chunchun-Dong, Jangan-Gu, Suwon City,
Kyunggi-Do 440-746, Korea
kwonho@live.co.kr, chkimbio@skku.edu

Abstract. Based on the sequences of various CMAH genes, our group previously cloned the pig CMAH gene (*pcmah*) that encodes the CMP-N-acetylneuraminic acid hydroxylase from pig tissues. In this study, we have cloned the complete *pcmah* gene by using rapid amplification of cDNA ends (RACE), and also isolated 5' alternative transcription start sites and 3' alternative splicing variants of *pcmah*. 5'RACE analysis revealed that *pcmah* has two alternative spliced forms, with two different initiation sites located in exon 0 and exon 1a of the gene expressed by pig small intestine. Also, 3' RACE revealed that *pcmah* has three 3' alternative splicing variants, *pcmah* variant 2 with nine exons and variant3 eight exons. Furthermore, the mRNA expression of the *pcmah* gene and its variants were analyzed in various pig tissues by reverse transcription polymerase chain reaction (RT-PCR).

1 Introduction

"Sialic acid" is a general term indicating N-acylneuraminic acids and their many derivatives (Schauer et al., 1997; Irie and Suzuki 1998). N-acetylneuraminic acid (NeuAc) and N-glycolylneuraminic acid (NeuGc) are two of the most common sialic acid types. NeuAc is expressed ubiquitously and is the major sialic acid form found in humans. NeuGc is abundant in most mammals, but not in humans due to a homozygous deletion mutation in the enzyme responsible for NeuGc biosynthesis (Chou et al., 1998). However, NeuGc is produced from NeuAc in humans, through a reaction catalyzed by a specific hydroxylase, cytidine-5'-monophospho-N-acetylneuraminic acid hydroxylase (CMAH), that converts CMP-NeuAc to CMP-NeuGc (Fig. 1) (Kawano et al., 1995; Kozutsumi et al., 1990; Muchmore et al., 1989; Shaw and Schauer, 1988).

---

* Corresponding author.

K.-H. Song and C.-H. Kim: *Sialo-Xenoantigenic Glycobiology*, SKKU 1, pp. 11–18.
DOI: 10.1007/978-3-642-34094-9_2   © Springer-Verlag Berlin Heidelberg 2013

Although CMAH has been cloned and characterized in various animals (Kawano et al., 1995; Martensen et al., 2001; Schlenzka et al., 1996; Schlenzka et al., 1994), the pig CMAH gene has only been partially cloned so far (Schlenzka et al., 1996). The amino acid sequences of the mouse, pig, and chimpanzee CMAH enzymes have been found to be highly homologous (Chou et al., 1998; Kawano et al., 1995; Schlenzka et al., 1996). Therefore, based on the sequences of various CMAH genes, our group previously cloned the pig CMAH gene (*pcmah*) that encodes the CMP-*N*-acetylneuraminic acid hydroxylase from pig tissues. In this study, we have cloned the complete *pcmah* gene by using rapid amplification of cDNA ends (RACE), and also isolated 5' alternative transcription start sites and 3' alternative splicing variants of *pcmah*. Furthermore, the mRNA expression of the *pcmah* gene and its variants were analyzed in various pig tissues by reverse transcription polymerase chain reaction (RT-PCR).

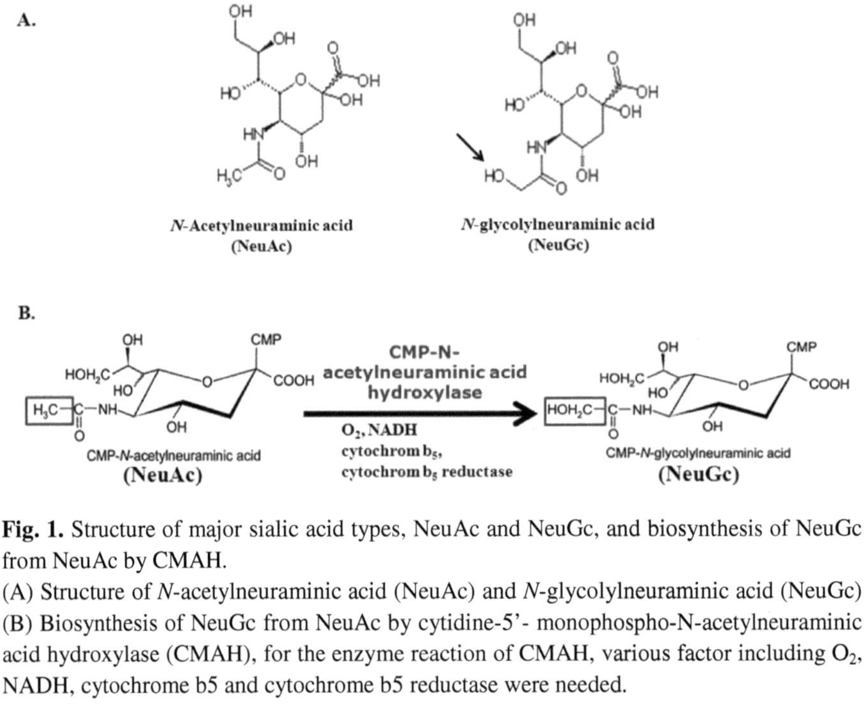

A.

B.

Fig. 1. Structure of major sialic acid types, NeuAc and NeuGc, and biosynthesis of NeuGc from NeuAc by CMAH.
(A) Structure of *N*-acetylneuraminic acid (NeuAc) and *N*-glycolylneuraminic acid (NeuGc)
(B) Biosynthesis of NeuGc from NeuAc by cytidine-5'- monophospho-N-acetylneuraminic acid hydroxylase (CMAH), for the enzyme reaction of CMAH, various factor including O_2, NADH, cytochrome b5 and cytochrome b5 reductase were needed.

2 Materials and Methods

2.1 *Rapid Amplification of cDNA Ends (RACE)*

The RACE reaction was performed using the GeneRacer[TM] Kit (Invitrogen). The total RNA from pig small intestine was used as a template for RACE with each primer set in Table 1. To obtain the 5' ends, the first strand cDNA was synthesized by reverse transcriptase with a 5' gene specific primer (5'GSP1) using SuperScript[TM] III Reverse Transcriptase (Invitrogen). The first and second PCR were performed using the following respective primer sets: GeneRacer[TM] 5'primer and 5'GSP2, GeneRacer[TM] 5'nested primer and 5'GSP3. To obtain the 3' ends, the first strand cDNA was amplified using a GeneRacer Oligo dT primer. The first and second PCR were performed using the following respective primer sets: GeneRacer[TM] 3'primer and 3'GSP1, GeneRacer[TM] 3'nested primer and 3'GSP2. The amplified 5'- and 3'-RACE products were subcloned and sequenced.

2.2 Reverse Transcription Polymerase Chain Reaction (RT-PCR)

cDNA from various pig tissues was amplified by PCR with each primer set in Table 1 (for *pcmah*, pCMAH-S and pCMAH-exon3-AS; for 5'UTR-1, UTR1-S and UTR-AS; for 5'UTR-2, UTR2-S and UTR-AS; for Full, pCMAH-S and pCMAH-full-AS; for V2, pCMAH-S and pCMAH-V2-AS; for V3, pCMAH-S and pCMAH-V3-AS) using EF-Taq polymerase (SolGent, Korea). The use of equal amounts of mRNA in the RT-PCR assay was confirmed by analyzing β-actin expression levels. The PCR products were separated by a 1% agarose gel, stained with EtBr, and visualized by UV.

Table 1. Primers used for gene RACE and RT-PCR

Primers for 5'RACE	
5'GSP1	AACCACCATCCTCGCGCAAAAGC
5'GSP2	GCGTGAGTAAG GTACGTGATCTGC
5'GSP3	TCGTCTTGACAGAAGCTTCCAGGA
GeneRacerTM 5'	CGACTGGAGCACGAGGACCATGA
GeneRacerTM 5'nested	GGACACTGACATGGACTGAAGGAGTA
Primers for 3'RACE	
3'GSP1	GACCACCTGAGTTACCCAACACTG
3'GSP2	TGGCAACACGGAAAGACCTGTATT
GeneRacerTM 3'	GCTGTCAACGATACGCTACGTAACG
GeneRacerTM 3'nested	CGCTACGTAACGGCATGACAGTG
Primers for RT-PCR	
pCMAH-S	ACGGTACCATGAGCAGCATCGA
pCMAH-exon3-AS	ACAACCAGTTCGTCTTGACAG
pCMAH-full-AS	AACTCGAGCCCAGAGCACATCA
pCMAH-V2-AS	ACCTCGAGTATCCAGCCATACT
pCMAH-V3-AS	ACCTCGAGATTATCTTTGTATT
UTR1-S	GTCAACGGAAATACTGAGCTGGGT
UTR2-S	AGGCACTGTTATTGGTGCCTCCTG
UTR-AS	TCGTCTTGACAGAAGCTTCCAGGA

3 Results

3.1 Isolation of pcmah and Its Alternative Transcript

Previously, a *pcmah*-1 fragment of 1734 bp was isolated by PCR from pig small intestine RNA using a degenerative forward primer (5'dgP) and gene-specific reverse primer (3'GSP) (Fig. 2) (Kang, 2007). To isolate the complete coding sequence and transcription start site of *pcmah*, we performed 5'RACE and finally obtained two specific products of 527 bp (5'*pcmah*-1) and 626 bp (5'*pcmah*-2) (Fig. 2).

DNA sequence analysis using these three fragments, *pcmah*-1, 5'*pcmah*-1 and -2, established that the complete *pcmah* gene contains a 1,734 bp ORF encoding 577 amino acids and two 5' UTR sequences, 5'UTR-1 (228 bp) and 5'UTR-2 (327 bp) (Fig. 3A). These cDNA sequences have been deposited in GenBank (GenBank accession number EU204974, FJ907456 and FJ907457). A BLASTn analysis of GenBank was performed to understand the genomic structural organization of the *pcmah* gene. As shown in Table 2 and Fig. 3A, the *pcmah* gene has two alternative spliced forms. The longer form is 76,863 bp long with a 5'UTR of 327 bp (5'UTR-2) and fifteen exons starting with exons 0 and 1b. The shorter form is 36,241 bp long with a 5'UTR of 228 bp (5'UTR-1) and fourteen exons starting with exon 1a. The 3' ends of 5'UTR-1 and 5'UTR-2 overlap for 103 bp (Fig. 3). These results indicated that alternative spliced forms of the 5' UTR of *pcmah* does not effect on the ORF sequence of *pcmah* (Fig. 4).

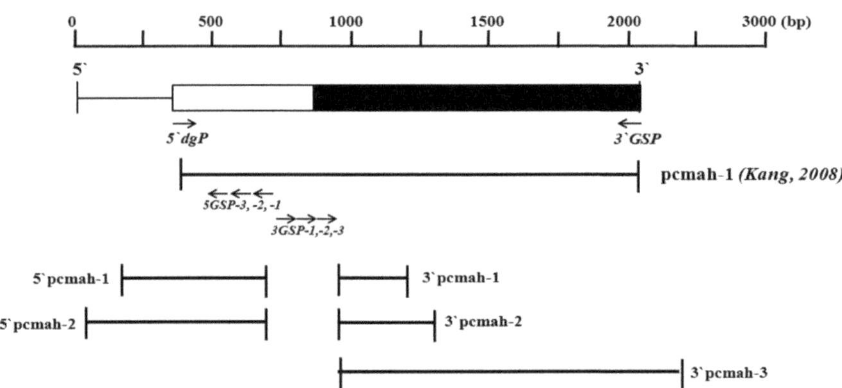

Fig. 2. Cloning strategy for cDNA encoding pCMAH.
The coding region for *pcmah* depicted as a box. A shade box indicates region of pig partial CMAH gene previously cloned by Schlenzka,W. et al., and an open box indicates newly isolated region by Kang in previous study. The fragments, 5'*pcmah*-1, -2, 3'*pcmah*-1, -2 and -3 were acquired by RACE. Arrows indicate primer sites for amplification of fragments or RACE.

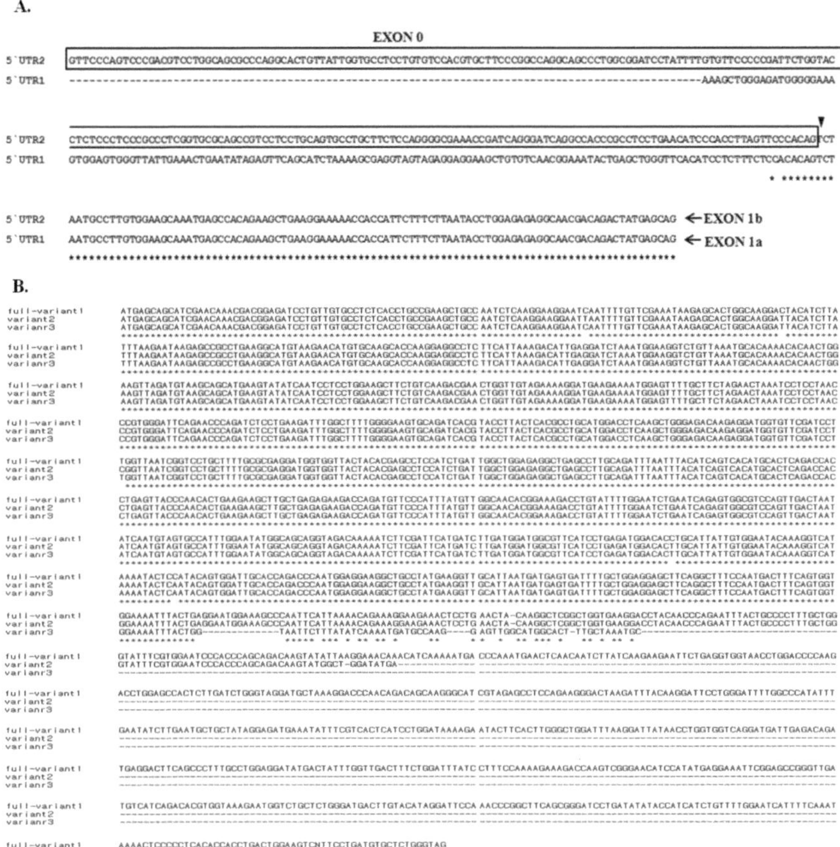

Fig. 3. Comparison of sequences of the *pcmah* 5' and 3`variants.
(A) Comparison of 5'UTR-1 and 2 of *pcmah*. The 5'UTR-1 is composed only of exon 1a, whereas the 5'UTR-2 is composed of exon 0 and exon 1. The box indicates the exon 0. The arrow indicates the translation start site, ATG. (B) Comparison of 3' alternative splicing variants of *pcmah*. (*): Fully conserved residues, (:): conservation of strong groups and (.): conservation of weak groups.

In addition, from sequence analysis of 3' RACE, three products suggesting 3' alternative splicing variants were obtained (Fig. 2). The ORF sequence of the longest form (3'*pcmah*-3, *pcmah* full) containing fourteen exons corresponds to *pcmah*-1, whereas the two shorter forms (3'*pcmah*-1 and 2) each consist of nine and eight exons (Fig. 2 and 4). These short forms were named *pcmah* variant 2 and 3, and the sequence information has been deposited in GenBank (GenBank accession number EU271934 and EU271935).

The deduced *pcmah* amino acid sequence has typical motifs such as a binding site for a rieske iron-sulfur center, postulated binding sites for a mononuclear iron

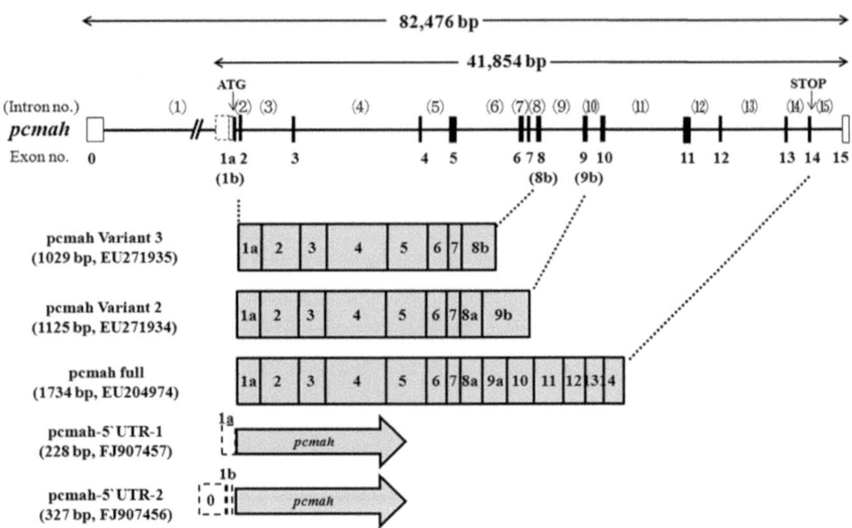

Fig. 4. Genomic structure of the *pcmah*.
Shade boxes: coding exons, open boxes: untranslated exons, solid lines: introns. The positions of the start codon ATG and the stop codon are indicated by arrows. Exon rearrangement of each variant was represented by numbered gray boxes.

Table 2. Exon/intron splice junctions of the *pcmah*.
a Exon length determined by sequencing.
b Intron length determined by sequence analysis using Blastn search of GeneBank.

Exon no.	Exon length (bp)[a]	Exon sequence		Intron no.	Intron length (bp)[b]	Intron sequence	
		5` Junction	3` Junction			5` Junction	3` Junction
0	237	GTTCCC .... CCACAG		1	37849	gtgaga .... gctgcc	
1a	98	TCTAAT .... GAGCAG		2	237	gcaagt .... tcatag	
1b	236	AAAGCT .... GAGCAG		-	-	- - -	
2	204	CATCGA .... TGGAAG		3	4254	gtactg .... atccag	
3	92	GTCTGT .... AACTGG		4	9165	gtaaat .... ttctag	
4	110	TTGTAG .... GTGCAG		5	1615	gtaagg .... caacag	
5	191	ATCACG .... CCTGAG		6	3666	gtaagg .... ccctag	
6	148	TTACCC .... CAGCAG		7	300	gtctgt .... ttccag	
7	82	GTAGAC .... ACAAAG		8	738	gtattt .... ccacag	
8	138	GTCATA .... TTACTG		9	2430	gtaatt .... ttccag	
9	136	AGGAAT .... AGACAA		10	457	gtatgg .... tgtcag	
10	141	GTATAT .... AGACAG		11	4079	gtttga .... tttaag	
11	175	CAAGGG .... GTCAGG		12	2282	gtatgc .... gaacag	
12	117	ATGATT .... GAGGAA		13	5039	gtaagc .... tttcag	
13	118	ATTCGG .... ATCTGT		14	2483	aagtcc .... acaggt	
14	74	TTTGGA .... GGGTAG				....	

center, a possible CMP-NeuAc binding site, and possible site of interaction with cytochrome b5, which are highly conserved with other known sequences with the exception of human (Martensen et al., 2001). A multiple alignment analysis among the amino acid sequences of alternative splicing variants was performed. *pcmah* full contains all of the functional motifs, whereas *pcmah* variant 2 lacks the second postulated binding site for a mononuclear iron center and *pcmah* variant 3 lacks the second postulated binding site for a mononuclear iron center as well as the possible site of interaction with cytochrome b5.

3.2 Expression Pattern of the pCMAH mRNAs in Various Pig Tissues

To investigate the expression patterns of the *pcmah* gene, total RNA was isolated from various pig tissues and RT-PCR analysis was performed with a primer set which amplifies exon 1 to exon 3. The *pcmah* gene was found to exhibit distinct tissue-specific expression patterns. The gene was most highly expressed in the small intestine, and moderately expressed in the rectum, tongue, spleen, and colon. However, *pcmah* full was minimally expressed in the brain, testicle, liver, bladder, stomach, muscle, kidney, spinal cord, and heart (Fig. 5). To investigate the tissue-specific expression of the alternative spliced forms of the 5'UTR, RT-PCR analysis was performed with primer sets specific for 5'UTR-1 or 5'UTR-2. These alternative spliced forms have different expression patterns in various pig tissues, except for the rectum. 5'UTR-1 was mainly expressed in the small intestine and colon. However, the 5'UTR-2 form was mainly expressed in the spleen, tongue, testicle, kidney, and liver. Furthermore, mRNA expression of 3' alternative splicing variants was investigated in the pig tissues which express *pcmah*. Variant 2 was weakly expressed in the small intestine and colon. Variant 3 was highly expressed in the rectum, spleen and small intestine, and weakly expressed in the colon.

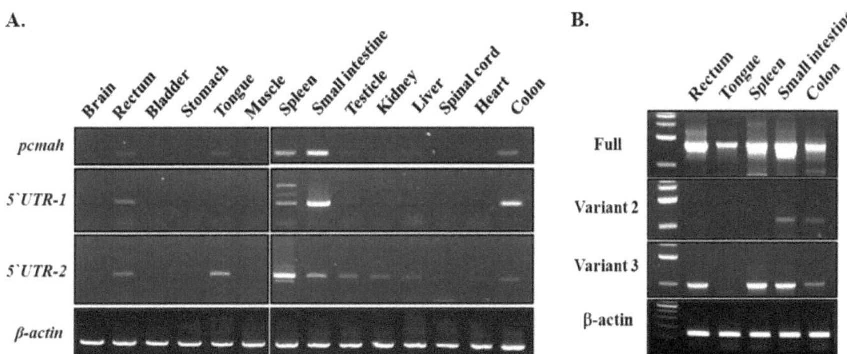

Fig. 5. mRNA expression of *pcmah* and its variants in various pig tissues.
(A) Relative expression levels of 5'UTR-1, -2 and mRNA of *pcmah* were measured by RT-PCR in various pig tissues. (B) Relative expression levels of alternative splicing variants of *pcmah* were measured by RT-PCR. *β*-Actin was used as a control.

4 Discussion

NeuGc is produced from NeuAc through enzymatic hydroxylation of the N-acetyl residue of free NeuAc, CMP-NeuAc, or glycoconjugate-linked NeuAc (Buscher et al., 1977; Schauer and Wember, 1971). Even though CMAH genes of several animal species have been successfully cloned (Kawano et al., 1995; Kozutsumi et al., 1990; Muchmore et al., 1989; Shaw and Schauer, 1988) the pig CMAH gene has not yet been completely cloned (Schlenzka, 1996). In this study, a full ORF of the pig CMAH gene was cloned by RT-PCR and 5' RACE based on the known partial sequence of the pig CMAH gene and homologous known CMAH genes. The cloned *pcmah* gene was found to share about 93% sequence homology with that of cattle, chimpanzee, and human, and 89% sequence homology with mouse. NeuGc expression is widespread in animals, and also shows tissue-specific and developmentally-regulated expression (Varki, 1992; Muchmore, 1992). Mouse CMAH mRNA is expressed in various tissues including the liver, thymus, spleen, and kidney, but not in the brain (Kawano, 1995). *pcmah* mRNA expression was investigated in several pig tissues. RT-PCR analysis revealed that *pcmah* mRNA is highly expressed in the small intestine and spleen, and moderately expressed in the rectum, tongue, testes, liver and colon. *pcmah* mRNA was not detected in the brain, bladder, stomach, muscle, pancreas, kidney, spinal cord, or heart. High hydroxylase activity, a large amount of the hydroxylase protein, and a high NeuGc/NeuAc ratio in the small intestine and spleen (Malykh et al., 1998) was found to be due to high levels of *pcmah* mRNA. NeuGc-containing glycoconjugates in the pig small intestine play an important role in mediating infections by pathogens such as *Escherichia coli K99* (Malykh et al., 2003).

5'RACE analysis revealed that *pcmah* has two alternative spliced forms, with two different initiation sites located in exon 0 and exon 1a of the gene expressed by pig small intestine. The shorter transcript was named 5'UTR-1 because in the pig small intestine, expression of the shorter transcript whose transcription is initiated in exon 1a (5'UTR-1) is dominant over the longer transcript whose transcription is initiated in exon 0 (5'UTR-2). The expression patterns of *pcmah*5'UTR-1 and -2 in various pig tissues were investigated using RT-PCR. 5'UTR-1 is mainly expressed in the small intestine and colon, whereas 5'UTR-2 is highly expressed in the spleen, and moderately expressed in most tissues expressing *pcmah*, including the rectum, tongue, small intestine, testicle, kidney, and colon. Extra multiple RT-PCR products observed from the RNA obtained from spleen tissue were determined to be artifacts by sequencing.

Also, 3' RACE revealed that *pcmah* has three 3' alternative splicing variants, *pcmah* variant 2 with nine exons and variant3 eight exons. Expression of these variants was investigated in pig tissues such as the rectum, tongue, spleen, small intestine, and colon, which are main sites of *pcmah* expression. Interestingly, the small intestine and colon expressed all three *pcmah* variants, whereas the tongue only expressed the *pcmah* full form. The rectum and spleen expressed *pcmah* full and variant 3, and not variant 2. The biological role of *pcmah* alternative splicing is not yet clear, and thus it will be useful to clarify the difference of NeuGc expression or CMAH activity in various pig tissues.

Chapter 3
Functional Characterization of pCMAH in the Synthesis of N-Glycolylneuraminic Acid as the Xenoantigenic Determinant in Pig-to-Human Xenotransplantation

Kwon-Ho Song and Cheorl-Ho Kim[*]

Molecular and Cellular Glycobiology Laboratory, Department of Biological Science,
SungKyunKwan University, 300 Chunchun-Dong, Jangan-Gu, Suwon City,
Kyunggi-Do 440-746, Korea
kwonho@live.co.kr, chkimbio@skku.edu

Abstract. To examine functional pCMAH activity, we analyzed the changes inNeuAc/NeuGc contents in *pcmah*-transfected PK15 and ECV304 cells. When human ECV304 cells negative in NeuGc expression were transfected with the *pcmah* cDNA, NeuGc contents were significantly increased in the transfectants compared to mock control cells. The transfected cells showed an increase in NeuGc contents and were capable of increasing human serum mediated cytotoxicity (HSMC) and xenoreactivity (HSMX).

1 Introduction

Glycoconjugate-bound NeuGc is the target epitope recognized by Hanganutziu-Deicher (HD) antibodies in immune reactions, first described independently by Hanganutziu and Deicher in the 1920s (Deicher, 1926 Hanganutziu, 1924). The HD antibodies appeared in patients after the therapeutic injection of animal antisera and were able to agglutinate animal erythrocytes (Higashi et al., 1977; Merrick et al. 1978). HD antibodies were also discovered in the sera of normal humans, as well as some disease-related patients (Nguyen et al., 2005; Yin et al., 2006). These findings indicate that NeuGc is a potential immunogen for humans. Recently, NeuGc was suggested to have the potential to be a non-Gal xenoantigen in pig-to-human xenotransplantation after α1,3GT is knocked out (Miwaet al., 2004; Morozumi et al., 1999). In this study, when *pcmah* was stably transfected into pig kidney PK15 cells and human endothelial ECV304 cells, the transfected cells showed an increase in NeuGc contents and were capable of increasing human serum mediated cytotoxicity (HSMC) and xenoreactivity (HSMX).

---

[*] Corresponding author.

K.-H. Song and C.-H. Kim: *Sialo-Xenoantigenic Glycobiology*, SKKU 1, pp. 19–26.
DOI: 10.1007/978-3-642-34094-9_3   © Springer-Verlag Berlin Heidelberg 2013

2 Materials and Methods

2.1 Cell Culture

The pig kidney cell line, PK15, was obtained from the Korean Cell Line Bank (Seoul, Korea) and cultured in Dulbecco's modified Eagle's medium (DMEM; WelGENE, Korea). ECV304 cells, an immortalized human vascular endothelial cell line, was obtained from American Type Culture Collection (Rockville, MD) and cultured in M199 medium (WelGENE, Korea). All culture media was supplemented with 10% heat-inactivated fetal bovine serum (FBS; WelGENE, Korea), 100 unit/ml penicillin, and 100 µg/ml streptomycin. The cultures were maintained in a 5% CO_2 atmosphere at 37°C.

2.2 Construction of the pcmah Expression Vector

The cDNA of the full pig CMAH ORF was inserted at the 3' end of the CMV promoter into the *Kpn* I and *Xho* I sites of pcDNA™3.1/myc-His expression vector (Invitrogen, Carlsbad, CA). To clone the full pig CMAH ORF, PCR was performed with pig cDNA as template and the following primer set containing restriction enzyme sites (*Kpn*I and *Xho*I): pig CMAH full 5'-ACGGTACCATG AGCAGCATCGA-3'   (sense),   5'-AACTCGAGCCCAGAGCACATCA-3' (antisense).

2.3 Establishment of Stable Transfectants

Transfection of *pcmah* into PK15 and ECV304 cells was performed using theWelFect-EX™PLUS Transfection Reagent (WelGENE, Korea). The cells were selected for stable integration of transfected DNA by changing to a media containing 0.6 mg/ml of G418 (A.G. Scientific, Inc.) for several days. To confirm the establishment of stable PK15 and ECV304 cell lines transfected with *pcmah*, Western blot analysis was performed.

2.4 Establishment of pCMAH-Silenced Stable PK15 Cell Lines by shRNA

pCMAH RNA interference stable PK15 cell lines were generated using the pSilencer™ 3.1-H1 puro vector (Ambion, Austin, TX). Four DNA oligonucleotides, containing a terminal *BamH* I or *Hind* III restriction site and linker sequence (TTCAAGAGA) that forms looped structures, were designed for knocking down pCMAH following the manufacturer's protocol, linking the 19-nucleotide sense and antisense sequences as follows: *pcmah* sh1-sense 5'-GATCCGCTGCCAATCTCAAGGAAGTTCAAGAGACTTCCTTGAGATTGGC AGCTTTTTTGGAAA-3',   *pcmah*   sh1-antisense   5'-AGCTTTTCCAAAAAAGCTGCCAATCTCAAGGAAGTCTCTTGAACTTCCT

TGAGATTGGCAGCG-3', *pcmah* sh2-sense 5'-GATCCGTTTACTGAGGAATGGAAAGTTCAAGAGACTTTCCATTCCTCAG TAAATTTTTTGGAAA-3'and *pcmah* sh2-antisense 5'-AGCTTTTCCAAAAAATTTACTGAGGAATGGAAAGTCTCTTGACTTTCCA TTCCTCAGTAAACG-3'. The oligonucleotides were annealed, and the resulting insert was subcloned into the *Bam*HI or *Hind*III site of pSilencer™3.1-H1 puro vector. The purified plasmid was transfected into PK15 cells by using theWelFect-EX™ PLUS Transfection Reagent (WelGENE, Daegu, Korea). Stable clones were selected by 2 μg/mL puromycin and screened for their ability to knockdown *pcmah* mRNA expression by RT-PCR.

2.5 HPLC Analysis

Total cell lysates were treated with 0.1 N HCl at 80°C for 1 h to release sialic acids from glycoconjugates. The sialic acid hydrolysates were labeled using a fluorescent labeling kit (DMB labeling kit, TAKARA). HPLC of the labeled sialic acids was performed using a Gemini 5 C18 column (4.6 × 250 mm) (Phenomenex). Elution was done with a mixture of methanol, acetonitrile, and water, 7/9/84(V/V/V), at a flow rate of 0.9 ml/min. The fluorescence (excitation at 373 nm, emission at 448 nm) was monitored using a fluoromonitor (RF-2000, Dionex). For all HPLC chromatograms, sialic acids were quantified by comparison using known quantities of DMB-NeuAc/Gc derivatives (NeuAc; TAKARA, NeuGc; Sigma) as a standard.

2.6 HPAEC-PAD Analysis

The NeuGc content of pCMAH sh-PK15 cells was determined by high performance anion exchange chromatography with pulsed amperometric detection (HPAEC-PAD) using the CarboPac PA-1 column-equipped chromatographic system DX 500 (Dionex) in combination with the Aminotrap columns, according to the modified method of Townsend et al.. The modification is as follows: eluent-1: 100 mM NaOH; eluent-2: 100 mM NaOH with 300 mM NaOAc. The system was eluted at a flow rate of 1 ml/min through a gradient pump module using the following gradients: linear gradient to 20% eluent-1 and 80% eluent-2 at 20min; linear gradient to 100% eluent-1 at 25 min; and column equilibration: 20 min at 100% eluent-1; linear gradient to 90% eluent-1 and 10% eluent-2 at the elution used 100 mM NaOH/100 mM NaOAc and the flow rate was 1 ml/min. 100 μg of protein was hydrolyzed in 0.1N HCl at 80°C for 1 h to release sialic acids from glycoconjugates. The sialic acid hydrolyzates were applied tothe chromatography.

2.7 *Lactate Dehydrogenase (LDH) Release Assay*

The LDH assay was performed using the CytoTox 96®Non-Radioactive Cytotoxicity Assay Kit (Promega). The PK15 or ECV304 transfectants with the construct were plated at 5 X 10^4 and 2 X 10^4cells in 96-well trays, respectively.

After 15 h of incubation, the wells were washed twice with serum-free media to remove the LDH and then incubated with 40% normal human serum (NHS) that had been diluted with media. The plate was incubated for two hour at 37°C and the released LDH was then measured.

2.8 FACS Analysis

Transfected cells were seeded 1 day prior to the assay. Cells were washed twice with phosphate buffered saline (PBS) and harvested by scraping. For the human IgG or IgM binding assay, the harvested cells were reacted to NHS. After 30 min incubation on ice, the cells were washed twice, and then incubated on ice for 30 min with fluorescein isothiocyanate (FITC)-conjugated goat anti human IgG and IgM (Zymed Laboratories). To analyze the amount of Gal antigen, the cells were reacted to FITC-conjugated GSIB4 lectin (Sigma) on ice for 30 min. In amelioration of by flow cytometry, the cells were also stained using 1 μg/ml propidium iodide (PI) (BD Pharmingen) for 10 min in the dark at room temperature. The stained cells were washed twice and analyzed using a FACSCalibur (Becton Dickinson).

2.9 Western Blot Analysis

Anti-c-Myc (Santa Cruz) and goat anti-mouse IgG-HRP (Santa Cruz) were purchased for this study. Cells were harvested by scraping and washed twice with PBS and resuspended in RIPA lysis buffer (20 mM Tris pH 7.5, 150 mM NaCl, 1% Triton X-100, 2 mM EDTA, 10% glycerol, 0.1% SDS, 0.5% deoxycholate) containing protease inhibitor cocktail (1mM Na_3VO_4, 20 g/ml PMSF, 10 g/l leupepton, 50 mM NaF). Whole lysates from transfectants were separated by 12% SDS PAGE, transferred onto a nitrocellulose membrane and immunoblotted with antibodies. Bands were visualized by the ECL system.

3 Results

3.1 Induced Constitutive Expression of NeuGc in pcmah-Transfected PK15 Cells

We previously cloned three alternative splicing variants of *pcmah*. As shown in Table 1, *pcmah* full has all functional motifs, whereas *pcmah* variant 2 lacks the second postulated binding site for a mononuclear iron center and *pcmah* variant 3 lacks the second postulated binding site for a mononuclear iron center and possible site of interaction with cytochrome b5. Therefore, *pcmah* full was used to functionally characterize *pcmah*.

To test whether ectopically expressed *pcmah* can synthesize NeuGc, we previously established *pcmah*-transfected PK15 cells (Fig. 1) and their human serum mediated cytotoxicity (HSMC) was increased (Kang, 2008). Next, we measured sialic acid contents in *pcmah*-transfected PK15 cells using HPLC

Table 1. Functional motifs each pCMAH variant gene has

Functional motif	pCMAH Full	pCMAH V2	pCMAH V3
binding site of the Rieske iron-sulfur centre	O	O	O
postulated binding sites of a mononuclear iron centre	O	O	O
possible CMP-Neu5Ac binding site	O	O	O
possible site of interaction with cytochrome b5	O	O	X
postulated binding sites of a mononuclear iron centre	O	X	X

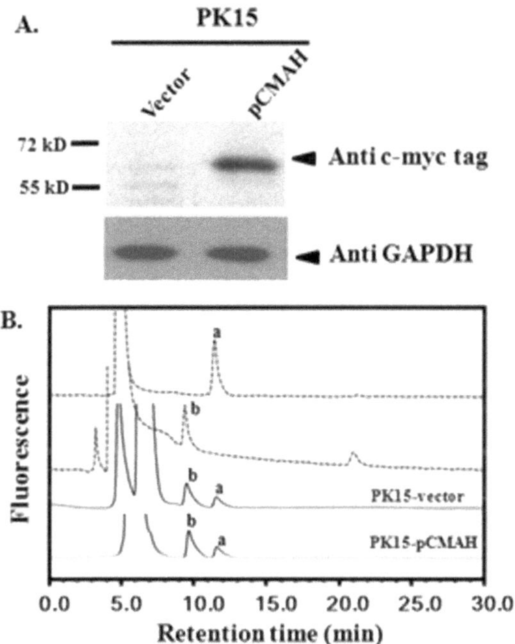

Fig. 1. Increased NeuGc in *pcmah*-transfected PK15 cells.
(A) Western blot analysis of vector and *pcmah*-transfected PK15 cells was performed by using an anti-c-myc antibody. (B) HPLC analysis of total sialic acid contents of vector and *pcmah*-transfected PK15 cells: a, NeuAc; b, NeuGc. The peak appearing after retention time 20 min is an artifact of the derivatization reagent. Each peak represents following area mean value: 'a' of dotted line indicating the standard NeuAc, 0.18; upper 'b' of dotted line indicating the standard NeuGc, 0.043; 'a' of PK15-vector line, 0.024; 'b' of PK15-vector line, 0.047; 'a' of PK15-pCMAH line, 0.06; 'b' of PK15-pCMAH line, 0.14.

analysis. HPLC analysis demonstrated that the *pcmah*-transfected PK15 cells contain a larger amount of NeuGc (70% of total sialic acids) than the vector-transfected cells (66% of total sialic acids) (Fig.1). These results indicate that NeuGc biosynthesis of *pcmah*-transfected PK15 cells was enhanced by *de novo* synthesis, reflecting activity of ectopically expressed *pcmah*. These results suggest that NeuGc generated by *pcmah* in PK15 cells has potential xenoantigenicity.

3.2 *Silencing of pCMAH in PK15 Cells*

The functional activity of pCMAH in PK15 cells was further characterized using shRNAs to silence pCMAH expression. PK15 cells were stably transfected with pCMAH-sh1, -sh2 and empty vector. As shown in Fig. 2A, *pcmah* mRNA expression was successfully down-regulated in the pCMAH-sh1 and -sh2 cells compared with psilencer 3.1 vector transfectants. Among the two pCMAH-sh cells, pCMAH-sh2 was more effective at silencing than pCMAH-sh1, and thus the pCMAH-sh2 cells were further analyzed. When the NeuGc contents of pCMAH-sh2 and psilencer 3.1-transfectants were analyzed using HPAEC-PAD, the NeuGc content of the pCMAH sh2 cells was less than that of the psilencer3.1-transfected cells (Fig. 2B). In addition, when HSMC was also determined by the LDH assay using 20% NHS, HSMC was decreased in the pCMAH-sh2 cells compared to the control, as expected (Fig. 2C). Therefore, the results strongly support that NeuGc generated by *pcmah* in PK15 cells has potential xenoantigenicity.

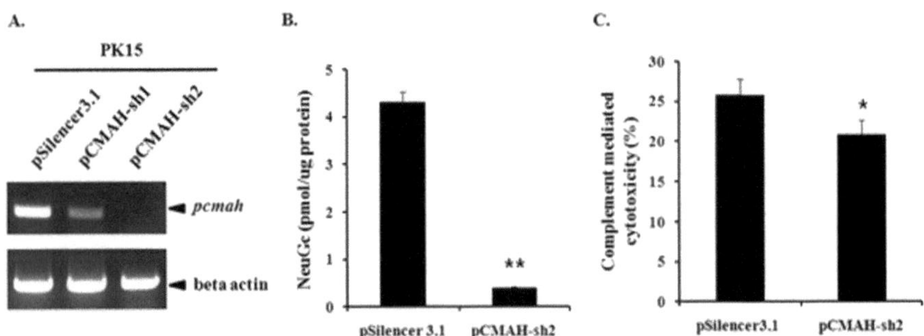

Fig. 2. The silencing effects of *cmah* expression using shRNA in PK15.
(A) RT-PCRanalysis of pCMAH-sh1, -sh2 and pSilencer3.1 vector-transfected PK15 cells was performed. (B) HPAEC-PAD analysis of the NeuGc contents of pCMAH-sh2 and pSilencer3.1-transfected PK15 cells. (C) HSMC of pCMAH-sh2 and pSilencer3.1 vector-transfected PK15 cells. NHS 20% was used as a source of the xenoreactive antibody and complement factors. Each value is expressed as the mean ± SD for three independent experiments. Differences in NeuGc amount and cytotoxicity were statistically tested using the Student t-test: *, $p < 0.05$; **, $p < 0.01$.

3.3 Overexpression of pcmah Inhuman Endothelial ECV304 Cells

Next, we investigated whether *pcmah* acts on human cells which do not normally express NeuGc. *pcmah*-transfected ECV304 cells significantly expressed pCMAH mRNA and recombinant pCMAH protein (Fig. 3A). As shown in Fig. 3B, the NeuAc/Gc contents of these transfectants were analyzed using HPLC. Vector-transfected ECV304 cells did not express NeuGc, but *pcmah*-transfected ECV304 cells expressed large amount of NeuGc. To determine HSMC, the LDH assay was performed using 50% NHS. As shown in Fig. 3C, HSMC was significantly greater in the *pcmah*-transfected ECV304 cells compared tothe control. These results demonstrate that the immunogenic NeuGc of *pcmah*-transfected ECV304 cells was supplied by *de novo* synthesis, which would reflect the activity of the pCMAH enzyme within the ECV304 cells, as also observed in PK15 cells.

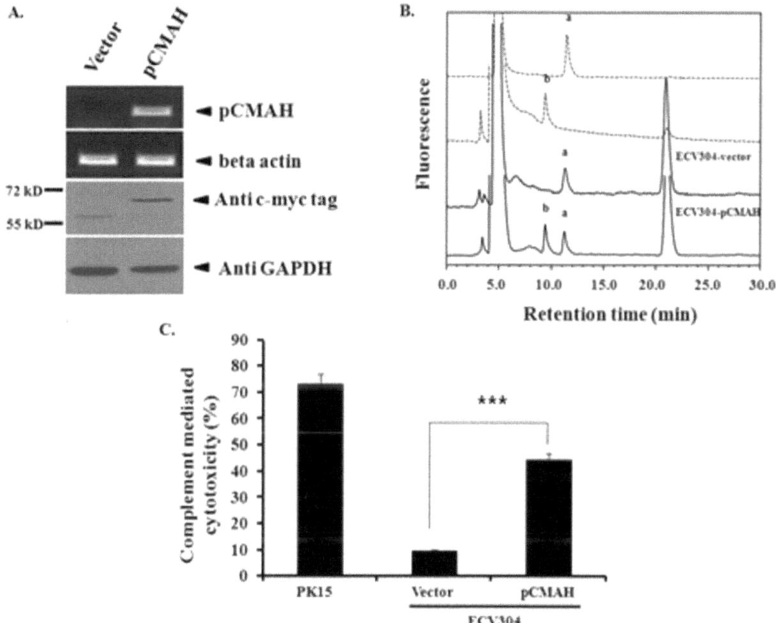

Fig. 3. NeuGc content and HSMC of *pcmah*-transfected ECV304 cells.
(A) RT-PCR and Western blot analyses of vector and *pcmah*-transfected ECV304 cells were performed. (B) HPLC analysis of the total sialic acid contents of the vector and *pcmah*-transfected ECV304 cells. a: NeuAc, b: NeuGc. The peak appearing after retention time 20 min is an artifact of the derivatization reagent. (C) HSMC of *pcmah*-transfected ECV304 cells. NHS 50% was used as a source of the xenoreactive antibody and complement factors. Each value was expressed as the mean ± SD for three independent experiments. Differences in cytotoxicity were statistically tested using the Student t-test: ***, p < 0.001.

4 Discussion

To examine functional pCMAH activity, we analyzed the changes inNeuAc/NeuGc contents in *pcmah*-transfected PK15 and ECV304 cells. When human ECV304 cells negative in NeuGc expression were transfected with the *pcmah* cDNA, NeuGc contents were significantly increased in the transfectants compared to mock control cells (Fig. 3B). However, the NeuGc contents of the *pcmah*-transfected PK15 cells were only slightly increased compared to themock transfectants (Fig. 1B), possibly due to the constitutive level of endogenous NeuGc in PK15 cells. CMAH expression is known to be a dominant factor in the production of glycoconjugate-bound NeuGc (Malykh et al., 1998) and expression of NeuGc-glycoconjugates is regulated by various factors including post-translational modification of the enzyme, and regulation of cytochrome $b_5$orcytchrome b_5 reductase accessibility (Kawano et al., 1995; Kawashima et al., 1993). In this study, the increased NeuGc contents in the transfected PK15 and ECV304 cells are attributed to the increase in CMP-*N*-acetylneuraminic acid hydroxylase activity. These results confirm that the cloned cDNA encodes for pig CMAH.

The Gal antigen was reported to be the major antigen in pig-to-human xenotransplantation (Cooper et al., 1994), and genetic approaches to modify this carbohydrate-antigen have also progressed (Sandrin et al., 1995; Tanemura et al., 1997). However, even after the removal of the Gal antigen by knocking out the α-1, 3 galactosyltransferase, non-Gal antigens including the HD antigen, Thomsen-Fridenreich (T or TF), or Forssman antigen still caused xenoreactivity (Ezzelarab et al., 2005; Miwa et al., 2004). N-glycolylneuraminic acids (NeuGc), also called the Hanganutziu-Deicher (HD) antigen, are widely expressed on endothelial cells of all mammals except humans and are targets for non-Gal antibodies (Miwa et al., 2004). It was also demonstrated that transfection of human immune cells with mouse CMAH increases human serum-mediated cytotoxicity (Nguyen et al., 2005). When treated with 20% NHS, HSMC was slightly elevated in *pcmah* over-expressed PK15 cells, compared to thecontrol (Kang, 2008). Moreover, pCMAH silencing by shRNA resulted in reduction of NeuGc content and xenoantigenicity in PK15 cells (Fig. 2). In general, the PK15 cells express NeuGc as well as the Gal antigen, whereas human ECV304 cells do not express NeuGc orthe Gal antigen. Therefore, 20% NHS treatment had little effect on *pcmah* over-expressed ECV304 cells (data not shown). However, when treated with 50% NHS, human serum mediated cytotoxicity was elevated in *pcmah* over-expressed ECV304 cells, compared to the control (Fig. 3C). These results indicate that the cloned *pcmah* gene is capable of increasing NeuGc expression.

Chapter 4
Identification of Two Alternative Promoters of the Pig CMP-*N*-Acetylneuraminic Acid Hydroxylase Gene

Kwon-Ho Song and Cheorl-Ho Kim[*]

Molecular and Cellular Glycobiology Laboratory, Department of Biological Science,
SungKyunKwan University, 300 Chunchun-Dong, Jangan-Gu, Suwon City,
Kyunggi-Do 440-746, Korea
kwonho@live.co.kr, chkimbio@skku.edu

Abstract. We identify the two alternative promoter regions, and name the 5'flanking region of exon 1a the P1 promoter, and the 5'flanking region of exon 0 the P2 promoter. The P2 promoter is active in both PK15 pig kidney cells and IPI-21 pig small intestine cells, and appears to have a higher level of basal activity than the P1 promoter in both cell lines, suggesting that P2 is a housekeeping promoter. In contrast, the P1 promoter is highly active in IPI-2I cells compared to thePK15 cells, where the P1 promoter has only basal activity, suggestive of an intestine-specific promoter. Furthermore, our results demonstrate that the Sp1 transcription factor is akey regulatory element important for the regulation of *pcmah* expression.

1 Introduction

Sialic acid is a generic term used for neuraminic acid and its many derivatives (Angata, T. et al, 2002, Irie, A .et al 1998, Schauer, R. et al. 1997). N-glycolylneuraminic acid (NeuGc) is one of the most common sialic acid types found in most mammals except mosthumans (Chou, H.H., et al. 1998). NeuGc is closely linked to various diseases including cancer, infections, and immune rejection response during pig-to-human xenotransplantation (Byres, E. et al. 2008, Hedlund, M., et al. 2008, Nguyen, D.H. et al. 2005). NeuGc is known as a tumor-associated antigen in various human cancers including colon carcinoma, retinoblastoma, breast cancer and melanoma (Hedlund, M., et al. 2008, Higashi, H., et al. 1984, Hirabayashi, et al. 1987), and acts as a target receptor for pathogens such as *Escherichia coli*K99, and for bacterial toxins such as subtilase cytotoxin secreted by Shiga-toxigenic *E. coli* (Byres, E. et al. 2008, Kyogashima,

---

[*] Corresponding author.

K.-H. Song and C.-H. Kim: *Sialo-Xenoantigenic Glycobiology*, SKKU 1, pp. 27–42.
DOI: 10.1007/978-3-642-34094-9_4   © Springer-Verlag Berlin Heidelberg 2013

M. et al. 1989). In addition, NeuGc is a non-Gal xenoantigen that acts as a xenoantigen after the major Gal xenoantigen is eliminated by knock-out of α-1,3-galactosyltransferase activity in pig-to-human xenotransplantation (Nguyen, D.H. et al. 2005).

Biosynthesis of NeuGc is mediated by a specific hydroxylase, cytidine-5'-monophospho-N-acetylneuraminic acid hydroxylase (CMAH), that converts CMP-NeuAc to CMP-NeuGc (Kawano, T. et al. 1995, Muchmore, E.A. et al. 1989, Schlenzka, W. et al. 1996). The mouse CMAH enzyme and gene (*cmah*) have been characterized, and the formation of NeuGc was shown to be regulated by *cmah* transcription, which is tissue-dependent (Kawano, T. et al. 1995, Shaw, L. and Schauer, R. 1989). In addition, the appearance of NeuGc is influenced by various endotoxins and intestinal parasites. For example, lipopolysaccharide (LPS)-induced mouse B-cell activation reduces CMAH mRNA expression (Karlsson, N.G. et al. 2000, Naito, Y. et al. 2007, Portner, A. et al. 1993). Moreover, the regulation of NeuGc biosynthesis in the developing pig small intestine is directly correlated to hydroxylase activity and CMAH mRNA expression levels (Malykh, Y.N. et al. 2003). In humans, *cmah* has lost a single exon corresponding to exon 6 of mouse *cmah*, which abrogates NeuGc expression in humans (Chou, H.H. et al. 1998, Irie, A. et al. 1998). Interestingly, a recent study reported that *cmah* expression is increased and the incomplete CMAH protein synthesized in human adult stem cells (Nystedt, J. et al. 2010), even though human CMAH is enzymatically inactive (Chou, H.H. et al. 1998, Irie, A. et al. 1998). Although these reports indicate that *cmah* expression is generally regulated by developmental processes and infectious conditions, the molecular mechanism(s) for regulation of *cmah*, directly related to NeuGc biosynthesis, remains unknown in all organisms.

We have cloned the complete pig *cmah* (*pcmah*) gene and isolated two 5' alternative transcripts (5'untranslated region (UTR)-1 and -2), suggesting the existence of alternative promoters for the *pcmah*gene (Song, K.H. et al. 2010). In addition, the first transcript, 5'*pcmah*-1, which contains exon 1a and acommon open reading frame (ORF) region (exons 2–14), is intestine specific, whereas the second transcript, 5'*pcmah*-2, which contains exon 0, exon 1b and the common ORF region, is expressed in most pig tissues expressing *pcmah*. The tissue-specific expression pattern of *pcmah* alternative forms suggests that *pcmah* expression is regulated in a tissue-specific manner by utilization of alternative promoters. The biological role of the alternative splicing of *pcmah* is not yet clear, and thus it would be useful to clarify differences inNeuGc expression or CMAH activity in various pig tissues. Here, We identify the two alternative promoter regions, and name the 5'flanking region of exon 1a the P1 promoter, and the 5'flanking region of exon 0 the P2 promoter. We further demonstrate that the alternative promoter use is responsible for the cell-specific expression pattern of 5'*pcmah*-1 and -2. Finally, we demonstrate that Sp1 transcription factor binding sites are necessary to express *pcmah* in both the alternative promoters. This study furthers the understanding of the transcriptional regulation of *pcmah* and may therefore contribute to a better understanding of the role of NeuGc in infection, developmental processes, and tumorigenesis.

2 Materials and Methods

2.1 Cell Culture

The PK15 pig kidney cell line was obtained from the Korean Cell Line Bank (Seoul, Korea). Cells were cultured in Dulbecco's modified Eagle's medium (DMEM; WelGENE, Seoul, Korea). The IPI-2I pig small intestine cell line was obtained from the European Collection of Cell Culture (ECACC) and cultured in DMEM containing 0.024 IU/ml insulin and 4 mM glutamine. DMEM was also supplemented with 10% fetal bovine serum (FBS; WelGENE), 100 unit/ml of penicillin and 100 μg/ml of streptomycin. The cultures were maintained in a 5% CO_2 atmosphere at 37°C.

2.2 Reverse Transcription-Polymerase Chain Reaction (RT-PCR)

Total RNA was isolated from PK15 and IPI-2I cells using TRIZOL reagent (Invitrogen, Carlsbad, CA), and cDNAs were RT-synthesized using an oligo dT-adaptor primer and AccuPower® RT-PreMix (Bioneer, Daejeon, Korea). The cDNA from PK15 and IPI-2I cells were PCR-amplified with the following primers using EF-Taq polymerase (SolGent, Daejeon, Korea): 5'*pcmah*-1 sense 5'-GTCAACGGAAATACTGAGCTGGGT-3', 5'*pcmah*-2 sense 5'-TGCTTCTCCAGGGGCGAAACC-3', 5'*pcmah*-1/2 antisense 5'-TCGTCTTGACAGAA GCTTCCAGGA-3', *pcmah* sense 5'-ATGAGCAGCATCGAACAAACG-3', and *pcmah* antisense 5'-ACAACCAGTTCGTCTTGACAG-3'. The use of equal amounts of mRNA in the RT-PCR assay was confirmed by analyzing the β-actin expression levels.

2.3 Construction of Plasmids

Pig BAC clones containing the *pcmah* gene or cloned 5'-flanking DNA was kindly provided by the National Livestock Research Institute, RDA. Varying length fragments of 5'-flanking regions of *pcmah* were isolated by PCR using a pig BAC clone as the template with the primer sets indicated in Table1. The PCR products were subcloned into the *Bgl*II and *Hind*III sites of the pGL3-Basic vector (Promega, Madison, WI, USA). Multiple independent clones were isolated and confirmed by DNA sequencing.

2.4 Generation of Mutations in the pcmah Promoter

Site-directed mutagenesis was performed using theQuikChange XL Site-directed Mutagenesis kit (Stratagene, San Diego, CA, USA). The primers summarized in Table 8 were used to introduce mutations into the *pcmah* promoter constructs. The PCR conditions were 95°C for 1 min, followed by 18 cycles of 95°C for 50 sec, 60°C for 50 sec, and 68°C for 7 min, with a final extension at 68°C for 7 min.

The PCR products were digested with *Dpn*I at 37°C for 1 h and transformed into XL10-Gold ultracompetent cells (Stratagene). Mutations in multiple transcription factor binding sites were generated from single mutation constructs by repeating mutagenesis with a different primer set. All mutant constructs were confirmed by DNA sequencing.

2.5 Transfection and Luciferase Assay

Cells were seeded in wells of 12-well culture plates 1 day prior to the assay. Cells were co-transfected with 0.25 pmol of *pcmah* promoter-luciferase reporter constructs and 0.25 *u*g of *β*-galactosidase reporter plasmid using WelFect-EX™ PLUS Transfection Reagent (WelGENE). After 24 h, cells were harvested and lysed with 1X Passive lysis buffer (Promega). Luciferase activity and *β*-galactosidase activity was assayed by the luciferase and *β*-galactosidase enzyme assay system, respectively (Promega). Luciferase activity was normalized tothe *β*-galactosidase activity in the cell lysate and calculated as an average of three independent experiments.

2.6 Electrophoretic Mobility Shift Assays (EMSA)

Single-stranded oligonucleotides for EMSA were synthesized commercially (Integrated DNA Technologies, Coralville, IA, USA). Sequences of the synthesized DNA probes are listed in Table 8. To generate the double-strand probes, complementary oligonucleotides were mixed together ata 1:1 molar ratio, heated to 95°C for 2 min, and cooled to 25°C for 45 min. Nuclear extract cells were prepared as described previously (Chung, T.W. et al. 2003). EMSA was performed using a gel shift assay system kit (Promega). Briefly, 1.75 pmol/*u*l of double-stranded oligonucleotide probe was end-labeled with [γ-^{32}P]ATP (3000 Ci/mmol; PerkinElmer, USA) using T4 polynucleotide kinase (Promega). Binding reaction mixtures containing 2 μg of nuclear extract and 2 μl of gel shift binding buffer [4% glycerol, 1 mM MgCl$_2$, 0.5 mM EDTA, 0.5 mM dithiothreitol (DTT), 50 mM NaCl, 10 mM Tris-HCl (pH 7.5), and 0.05 mg/ml poly(deoxyinosine-deoxycytosine)] with or without unlabeled wild-type or mutant double-stranded oligonucleotides for competition, were preincubated for 10 min at room temperature. The mixtures were incubated with the labeled probe for 20 min at room temperature. For super-shift assays, PK15 nuclear extract was incubated with 2 *u*g of anti-Sp1 (PEP2 Santa Cruz Biotechnology, Santa Cruz, CA, USA) overnight at 4°C in the presence of binding buffer before addition of the labeled probe. The sample mixtures were separated by electrophoresis using 4% nondenaturing polyacrylamide gel in 0.5×Tris-borate EDTA buffer at 250 V for 30 min. The gel was dried and detected by autoradiography.

Table 1. Primers and probes used for identification of pcmah promoter.
[a] restriction enzyme sites are underlined. [b] lowercase letters indicate mutated nucleotides.

Primers for 5' Deletion mutants[a]

P1(Hind III-AS)	ACAAGCTTCAGTATTTCCGTTGACACAG	P2(Hind III-AS)	ACAAGCTTGATCGGTTTCGCCCCTGG
P1-1600(Bgl II-S)	ACAGATCTTGAAGGACAGCTTTGGCTCT	P2-1546(Bgl II-S)	ACAGATCTGCTTCAATGAATCCCCCAG
P1-1100(Bgl II-S)	ACAGATCTAAGGATAACCTGACCCCTTTGC	P2-800(Bgl II-S)	CCAGATCTGTCAAATCCGAGCTACAGC
P1-700(Bgl II-S)	ACAGATCTTGAGGATGTGGGTTTGATCC	P2-432(Bgl II-S)	CCAGATCTAGCTAAGTTCCTGCTTCC
P1-542(Bgl II-S)	ACAGATCTTCGGATCTGACATTGCTG	P2-386(Bgl II-S)	ACAGATCTGCGCTGACACGAAGTCCG
P1-260(Bgl II-S)	ACAGATCTAAGTGTGTTGGCTTTGACCTG	P2-334(Bgl II-S)	AAAGATCTGCAGTGGGAGCAGCGGGC
P1-223(Bgl II-S)	ACAGATCTAGCTTGAACGGCTTAACCAAG	P2-315(Bgl II-S)	ACAGATCTAGGGCGCGGGAAGCACT
		P1-265(Bgl II-S)	ACAGATCTGAGCTCCCGGTGATGCA
		P2-210(Bgl II-S)	ACAGATCTCACGGGAAGGAACTGTTC

Primers for Direct mutagenesis[b]

P1-Sp1a MuF	TTGACCTGaaaGTaGGGGCTGGGTAGCTTGAAC	P2-Sp1a MuF	AGCAGCGaatttGGCGCGGGAAGCACTGGGGAT
P1-Sp1a MuR	TACCCAGCCCCtACtttCAGGTCAAAGCCAACA	P2-Sp1a MuR	TTCCCGCGCCaaattCGCTGCTCCCACTGCTC
P1-Sp1b MuF	GGTGGGGGaTaaaTAGCTTGAACGGCTTAACC	P2-Sp1b MuF	GATGCAGGaattcGGAGCCTTTTCCGAAGTAGC
P1-Sp1b MuR	TCAAGCTAtttAtCCCCCACCCCCAGGTCAAAG	P2-Sp1b MuR	GCTCCgaattCCTGCATCACCGGGAGCTCG
P1-Sp1ab MuF	TTGACCTGGGaaTGaGaaCTGGGTAGCTTGAACG	P2-Sp1c MuF	GAGGGTGcTGGaCGtcCTCCCGGTGATGCAGGGG
P1-Sp1ab MuR	CTACCCAGttCtCAttCCCAGGTCAAAGCCAAC	P2-Sp1c MuR	CCGGGAGgaCGtCCAgCACCCTCCTGCCAACGCG

Probes for EMSA[b]

P1-Sp1	ACCTGGGGGTGGGGGCTGGGTAGCTT	P2-Sp1a	GAGCAGCGGGCAGGGCGCGGG
P1-Sp1 Mut	ACCTGGGGGTGGGGGaTaaaTAGCTT	P2-Sp1a Mut	GAGCAGCGaatttGGCGCGGG
		P2-Sp1b	GAGGGTGTTGGGCGAGCTCC
		P2-Sp1b Mut	GAGGGTGcTGGaCGtcCTCC
		P2-Sp1c	TGATGCAGGGGGGGAGGAGCC
		P2-Sp1c Mut	TGATGCAGGaattcGGAGC

3 Results

3.1 Expression of Two Alternative Transcripts, 5'pcmah-1 and -2 in Various Pig Cell Lines

Our previous study (Song, K.H. et al. 2010) isolated two 5'alternative splicing variants of *pcmah* and demonstrated their tissue-specific expression pattern, as illustrated in Fig. 1A. To investigate the relevance of the tissue-specific expression to cells originating from different tissues or organs, here we analyzed the mRNA levels of the two alternative transcripts, 5'*pcmah*-1 and -2, in pig kidney-derived PK15 cells and pig small intestine-derived IPI-2I cells. 5'*pcmah*-2 was expressed in both pig cell lines, whereas 5'*pcmah*-1 was expressed in IPI-2I cells only (Fig. 1B). This result was consistent with the previously described tissue-specific expression pattern of *pcmah*, with 5'*pcmah*-1 being intestine-specific while 5'*pcmah*-2 is expressed in most tissues expressing *pcmah* (Song, K.H. et al. 2010).

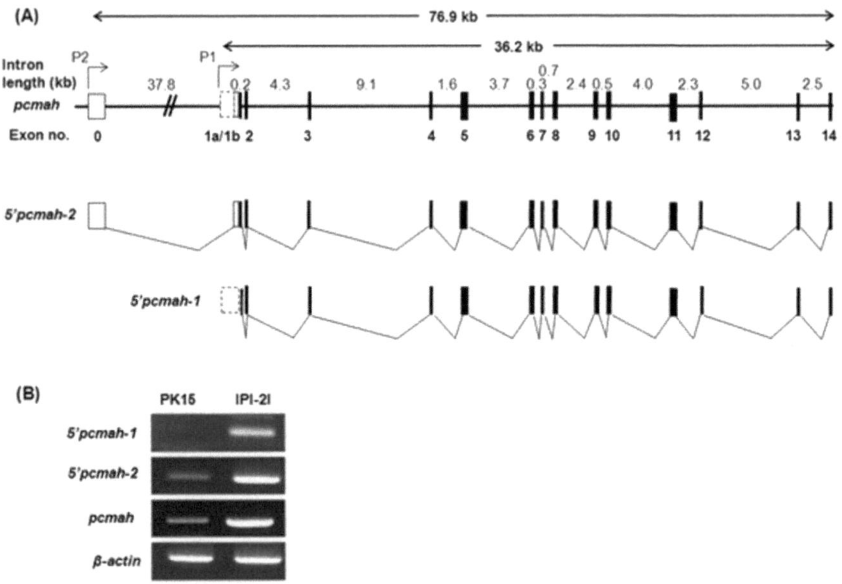

Fig. 1. Structure of *pcmah* and expression of 5'*pcmah*-1 and -2 in PK15 and IPI-2I cells
(A) Genomic structure of *pcmah* and two alternative transcripts. 5'*pcmah*-1 is composed of exon 1a and a common ORF region (exon 2–14). 5'*pcmah*-2 is composed of UTR regions containing exon 0 and exon 1b, and a common ORF region. Putative promoter regions, P1 and P2, are indicated by arrows. Shaded boxes indicate the coding exons and open boxes indicate untranslated exons. The dotted box indicates exon 1a. Solid lines represent introns and the numbers indicate the intron lengths. (B) Relative expression levels of 5'*pcmah* 1, 2 and *pcmah* (common region) were measured by RT-PCR in PK15 and IPI-2I cells. β-actin was used as a control.

3.2 Identification of Two Distinct *pcmah* Promoter Regions, P1 and P2

The different *pcmah* promoter regions – the 5' flanking region of exon 0 (promoter 2, P2) and the 5'flanking region of exon 1a (promoter 1, P1) – were PCR-amplified from the pig BAC clone containing the CMAH gene (Fig. 2). Three fragments of the P1 promoter (1600, 1100 and 700) and P2 promoter (1546, 800 and 432) were inserted into the pGL3-Basic vector and then transfected into PK15 and IPI-I2 cells. Promoter activity of these reporter constructs were determined by a luciferase reporter assay. The luciferase activity in PK15 and IPI-2I cells transfected with P1-1600 orP2-1546 fragments was 4–17-fold higher than activity in cells transfected with control pGL3-Basic empty vector (Fig. 2), suggesting that

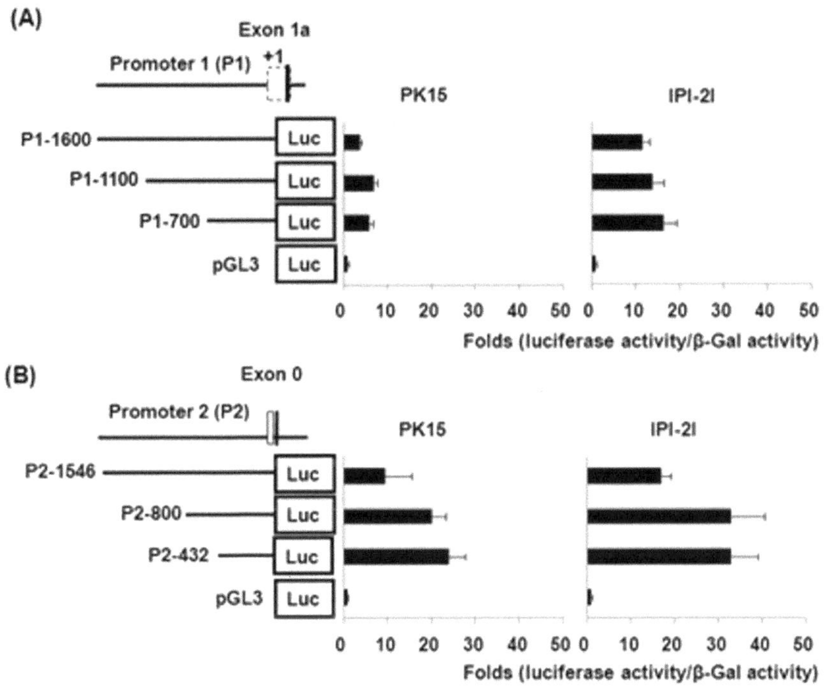

Fig. 2. Relative activities of the luciferase reporter plasmid with proximal region of exon 0 and exon 1a in PK15 and IPI-2I cells.
DNA fragments contained various lengths of P1 promoter region, which is a proximal region of exon 1a (A), and the P2 promoter region, which is a proximal region of exon 0 (B). The fragments were subcloned into a PGL3-basic vector (pGL3) and each of the resulting plasmids was transiently transfected into PK15 and IPI-2I cells. Luciferase and β-galactosidase activity in the transfected cells was measured. For each transfection, the luciferase activity was normalized with β-galactosidase activity and the relative fold value was determined from the ratio of the normalized activity and the activity in cells transfected with the empty pGL3-basic vector. Bars represent the mean±SE of three independent determinations.

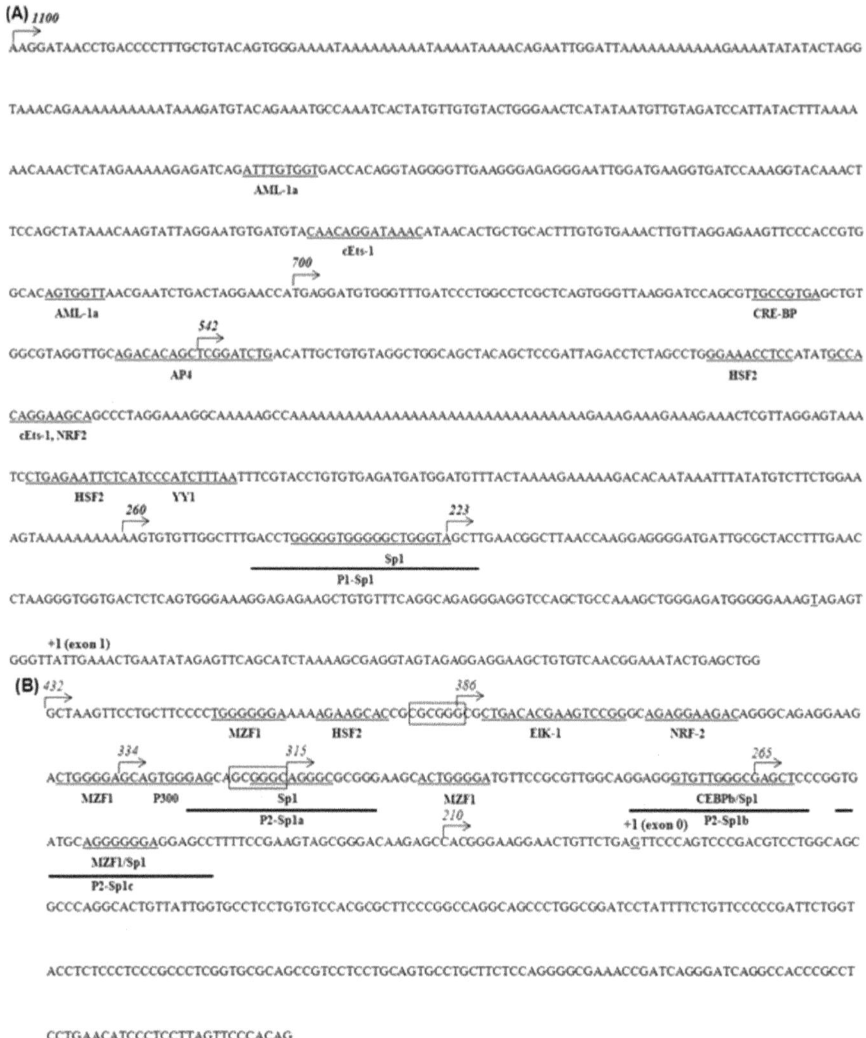

Fig. 3. Nucleotide sequences of the *pcmah* P1 and P2 promoter regions.
DNA sequences of P1-1100 (A) and P2-432 (B) fragments, which are basal promoter regions of *pcmah*, are represented. Putative binding sites for the transcription factors are underlined. Arrows indicate the end point of 5'deletion mutants. Boxes indicate GC box in (A). '+1'indicate the start of exon 0 (P2) or exon 1a (P1). Thick underlines indicate the oligonucleotide sequences used for EMSA.

the regionsdocontain promoter activity. In PK15 and IPI-2I cells, the overall activity of the P2 promoter was higher than the P1 promoter 1 (Fig. 2). Also, the maximum promoter activity of the P1 and P2 promoters were higher in IPI-I2 cells compared to PK15 cells (Fig. 2). These results were also consistentwith the cell-specific expression pattern of the 5'*pcmah*-1 and -2 transcripts (Fig. 1B).

We further investigated the basal promoter activity of the P1 and P2 promoters using PK15 cells. The highest activities were observed in the P1-1100 fragment for the P1 promoter and the P2-432 fragment for the P2 promoter. These fragments were further analyzed to characterize each promoter in PK15 cells. The nucleotide sequences of the two promoter regions of *pcmah* were analyzed (Fig. 3). The P1 and P2 promoters both lacked TATA and CCAAT boxes. Instead, four GC boxes were observed in the P2 promoter, whereas several polyA repeat sequences were evident in the P1 promoter. We further searched for potential transcriptional factor binding sites in these promoter regions using an algorithm (TFSEARCH). Through this analysis, we identified putative binding sites for AML-1a, c-Ets-1, CRE-BP, AP-4, HSF-2, NRF-2, YY1, and Sp1 in the *pcmah* P1 promoter (Fig. 3A). The *pcmah* P2 promoter contained putative binding sites for transcription factors including Mzf-1, HSF-2, Elk-1, NRF-2, p300, C/EBP, and Sp1 (Fig. 3B).

3.3 Characterization of Regulatory Elements of the pcmah Promoter

To determine the roles of the putative transcription factor binding sites that were identified in the two *pcmah* promoter regions, we generated several 5'-deletion constructs and analyzed their promoter activity in PK15 cells. Deletion of 419, 282 and 50 nucleotides from 1100, 542, and 260, respectively, decreased P1 promoter activity, suggesting that these regions may contain positive regulatory elements for P1 promoter activity. When the region was checked for transcription factor candidates using the algorithm mentioned above, binding sites for putative transcription factors including AML-1a, c-Ets-1, HSF2, NRF2, YY1, and Sp1 were found (Fig. 4A). Deletion of 42 and 52 nucleotides from the 432 and 386 fragments, respectively, increased P2 promoter activity, suggesting that the region between 432–334 fragments may contain putative regulators that negatively regulate promoter activity. However, deletion of 19, 50 and 55 nucleotides from the 334, 315, and 265 fragments, respectively, significantly decreased P2 promoter activities. There were putative binding sites for various transcription factors including two Mzf1, P300, C/EBP-β, and three Sp1 between the 334 and 210 fragments. These results indicate that one or more of these transcription factors may positively regulate *pcmah* transcription in PK15 cells (Fig. 4B). In addition, the P2 promoter contains two GC boxes: one located at the 391 point and the other overlapping with a Sp1a site. However, deletion of the 391 point did not decrease luciferase activity (Fig. 4B), but rather increased activity, indicating that the first GC box does not affect P2 promoter activity.

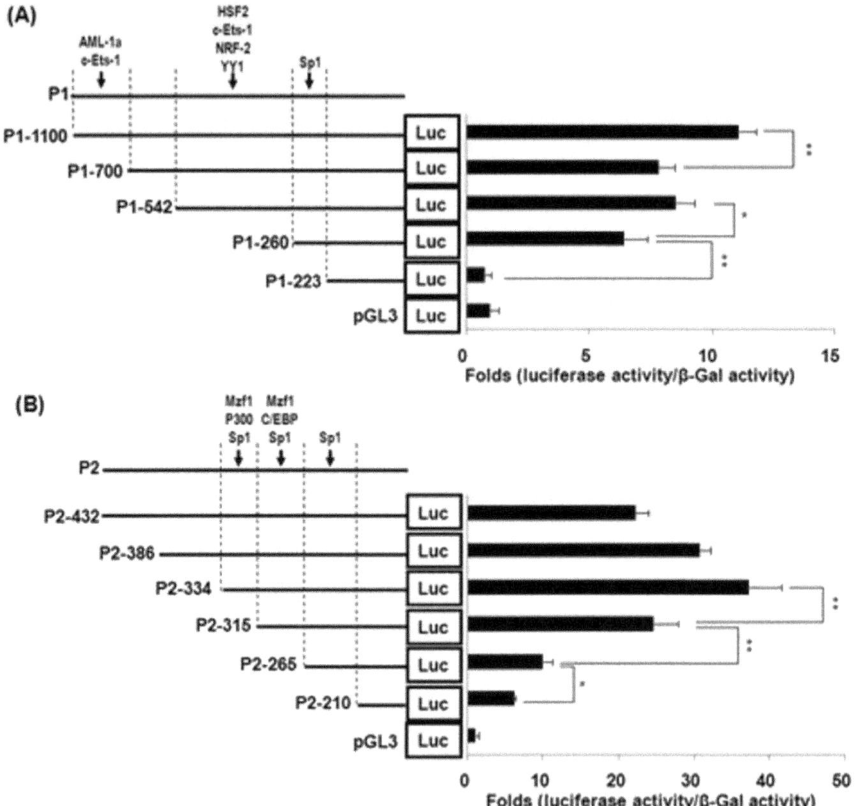

Fig. 4. Functional analysis of the *pcmah* P1 and P2 promoter regions.
The 5'deletion analysis of the P1 (A) and P2 (B) promoter regions of pig CMAH gene in PK 15 cells. DNA fragments of various lengths were subcloned into the PGL3-basic vector (pGL3) and each of the resultant plasmids was transiently transfected into PK15 cells. Luciferase and β-galactosidase activity in the transfected cells was measured. For each transfection, the luciferase activity was normalized with *β*-galactosidase activity and the relative fold value was determined from the ratio of the normalized activity and the activity in cells transfected with the empty pGL3-basic vector. Transcription factors existing in the region indicating black arrows are represented. Bars represent the mean±SE of three independent determinations. Differences in the fold value of luciferase activities were statistically tested using the Student t-test: *, $p < 0.05$; **, $p < 0.01$.

3.4 Sp1 Is Required for the Basal Promoter Activity of P1 and P2

There are three putative Sp1 binding sites in the P2-334 fragment and two putative Sp1 binding sites in the P1-260 fragment (Fig. 3). Deletion of the regions containing putative Sp1 binding sites markedly decreased luciferase activity (Fig. 5). Therefore, we further investigated the roles of the putative Sp1 binding sites in the *pcmah* P1 and P2 promoter regions. To examine whether Sp1 acts on

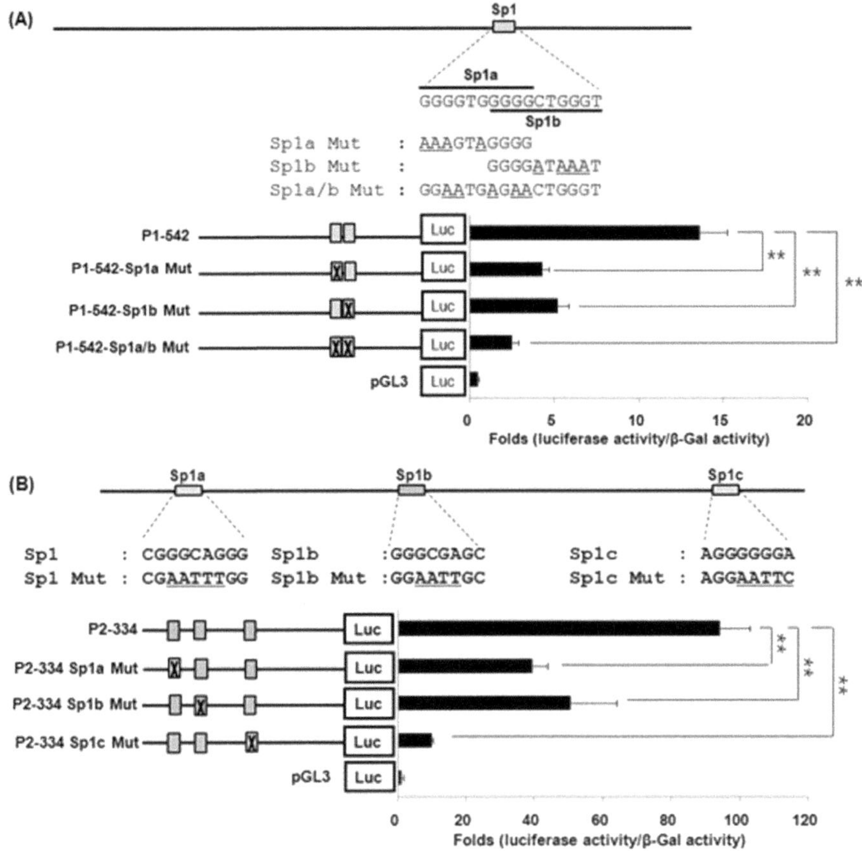

Fig. 5. Role of putative Sp1 binding sites on the luciferase activities of the P1 and P2 promoters.
(A) Nucleotide sequences of the wild type and mutant of two putative Sp1 binding sites (Sp1a and Sp1b) in P1-542 cells, and luciferase activities of three site-directed mutants and wild type P1-542 cells. (B) Nucleotide sequences of wild and mutant of three putative Sp1 binding sites (Sp1a, Sp1b and Sp1c) in P2-334 cells and luciferase activities of three site-directed mutants and wild type P2-334 cells. Mutated bases are underlined. Relative fold value was determined from the ratio of the normalized activity and the activity in the cell transfected with the empty pGL3-basic vector. Gray boxes indicate the putative Sp1 binding site. Bars represent the mean±SE of three independent determinations. Differences in the fold value of luciferase activities were statistically tested using the Student t-test: **, $p < 0.01$.

these regions as a transactivator, we altered the putative Sp1 binding sites by site-directed mutagenesis and examined the effect of each mutation on promoter activity in PK15cells. In the P2 promoter, mutation of Sp1a, Sp1b and Sp1c caused adecrease of approximately 68% (94- to 50-fold), 46% (94- to 51-fold), and 90% (94-to 10-fold), respectively, compared with the luciferase activity of wild-type P1-334 plasmid (Fig. 5). In the P1 promoter, luciferase activity of P1-542 Sp1a Mut, P1-542 Sp1b Mut, and P1-542 Sp1ab Mut, which ismutated in both Sp1a and Sp1b sites, wasreduced up to about 72% (14- to 4-fold), 65% (14- to 5-fold), and 79% (14-to 3-fold), respectively, compared to wild type P1-542 plasmid (Fig. 5). These results suggest that all three Sp1 binding sites of the P2 promoter and all two Sp1 binding sites of the P1 promoter affect *pcmah* promoter activity. To demonstrate that the Sp1 protein regulates *pcmah* promoter activity, we investigated the effects of a Sp1 inhibitor, mithramycin A, on the activity of each promoter. As expected, treatment with different concentrations (25–100 nM) of mithramycin A decreased the luciferase activity of both the P1 and P2 promoters in a dose-dependent manner (Fig. 6). Next, to confirm the direct interaction between Sp1 and the putative Sp1 binding sites in the P1 and P2 promoters, an EMSA assay was performed. PK15 cell nuclear extract was incubated with the ^{32}P-labeled oligonucleotide probes such as P1-Sp1, which containsthe two putative Sp1 binding sites overlapping in the P1 promoter region, orP2-Sp1a, P2-Sp1b, P2-Sp1c, which each contain one putative Sp1 binding site of the P2 promoter region. As shown in Figs. 7A and 7C, complexes that bound to each labeled oligonucleotide probe were detected. The specificity of binding was confirmed by a competition assay using 100-fold molar excess of each unlabeled oligonucleotide probe, and each unlabeled mutant probe carrying a point mutation in the Sp1 consensus sequence. As expected, the unlabeled probe efficiently competed for the binding of the transcription factor to each labeled oligonucleotide probe, whereas the unlabeled mutant probe did not compete (Figs. 7A and 7C). To further verify these results, a super-shift assay in the presence of anti-Sp1 was performed. Although the super-shift band was not observed in all cases (P1-Sp1, P2-Sp1a, P2-Sp1b and P2-Sp1c), the intensity of the major Sp1

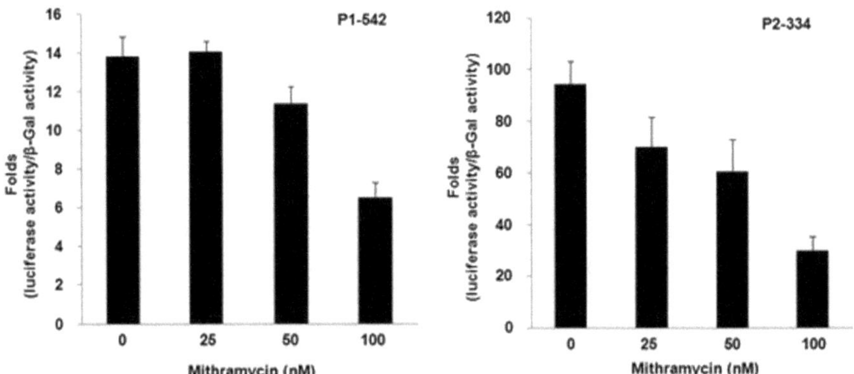

Fig. 6. Effects of the Sp1 inhibitor on P1 and P2 promoter activities.
P1-542 plasmid and P2-334 plasmid were transiently transfected into PK15 cells. The transfected cells were treated with 25, 50 or 100 nM) mithramycin A for 20 h. Luciferase and β-galactosidase activity in the transfected cells was measured. Relative fold value was determined from the ratio of the normalized activity and the activity in cells transfected with the empty pGL3-basic vector. Bars represent the mean±SE of three independent determinations.

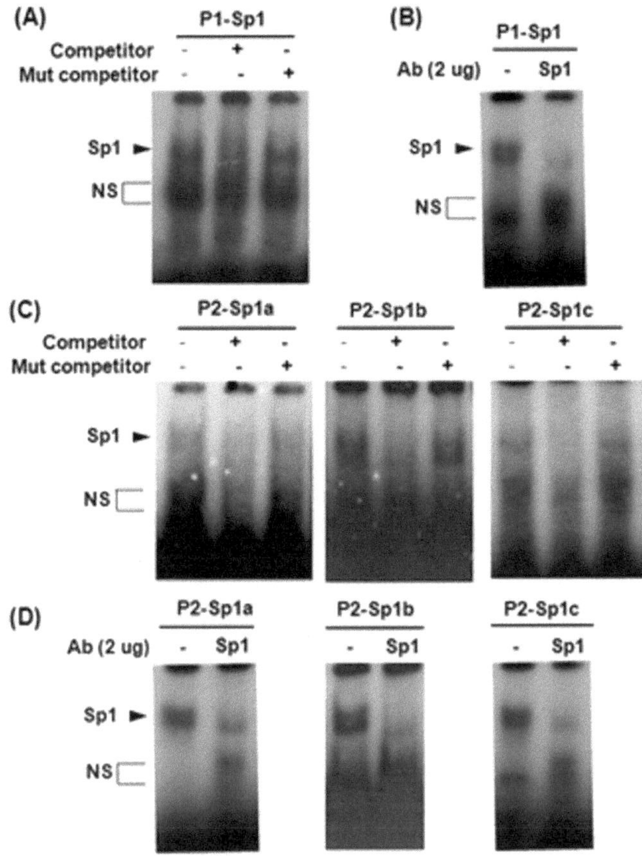

Fig. 7. Binding of Sp1 to binding sites in the P1 and P2 promoters.
PK15 cell nuclear extracts were incubated with ^{32}P labeled oligonucleotide probes including P1-Sp1 containing two overlapping Sp1 sites in the P1 promoter and in P2-Sp1a containing a Sp1a site in the P2 promoter, P2-Sp1b containing a Sp1b site in the P2 promoter or P2-Sp1c containing a Sp1c site in theP2 promoter. The mixture was subsequently analyzed by an EMSA. A competition assay was performed using a non-labeled probe or non-labeled mutant probe for P2-Sp1 of the P2 promoter (A), and P1-Sp1a, Sp1b and Sp1c of the P1 promoter (C). Super-shiftassay for the P2-Sp1 probe of P2 promoter (B) and for the probes including P1-Sp1a, Sp1b and Sp1c of P1 promoter (D) was performed using anti-Sp1 antibody. The arrowheads indicates specific binding complex for Sp1. NS: non-specific bands.

complex was markedly decreased by the addition of anti-Sp1 antibody, indicating competitive binding by Sp1 (Figs. 7B and 7D). These results indicate that Sp1 binds to the putative binding sites in the *pcmah* P1 and P2 promoter regions to positively regulate promoter activity.

4 Discussion

Several studies have provided direct or indirect evidence that NeuGc expression is tissue- and age-dependent, and may also be influenced by cell activation and disease status (Karlsson, N.G. et al. 2000, Malykh, Y.N. et al. 2003, Portner, A. et al. 1993). NeuGc formation can be mediated by two pathways: dietary incorporation of NeuGc (Tangvoranuntakul, P. et al. 2003) and biosynthesis by CMAH that converts CMP-NeuAc to CMP-NeuGc (Kawano, T. et al. 1995, Muchmore, E.A. et al. 1989, Schlenzka, W. et al. 1996). The former was evidenced by mechanistic results of NeuGc expression in some human fetal tissues and cancers (Devine, P.L. et al. 1991, Higashi, H. et al. 1985, Malykh, Y.N. et al. 2001, Varki, A. 2001). With regard to NeuGc production inthe latter case, expression of the CMAH enzyme is a dominant factor in the production of glycoconjugate-bound NeuGc (Malykh, Y.N. et al. 1998), and NeuGc expression is regulated by various factors including post-translational modification of the enzyme and regulation of cytochrome b_5 and cytchrome b_5 reductase accessibility (Kawano, T. et al. 1995). In mouse and humans, CMAH mRNA expression has been reported to be regulated during LPS-stimulated mouse B-cell activation and in human stem cells, as observed by RT-PCR (Naito, Y. et al. 2007, Nystedt, J., Anderson, H., et al. 2010). Although precise transcriptional approaches were not attempted due to limited information on *cis-* and *trans-*elements related to CMAH gene regulation in mouse and human, the mechanisms of transcriptional regulation should be interesting in terms of tissue-specific and environment-responsive expression. In the case of the pig, the level of CMAH mRNA in the developing jejunum has been correlated with variations in enzyme activity correlating with age and positional susceptibility for an enteric disease (Malykh, Y.N. et al. 2003). Although studies of transcriptional regulation should ultimately lead to understanding the relationship between CMAH mRNA expression and biological phenomena, no promoter studies were done to date on any of the *cmah* genes. Therefore, this study represents the first detailed characterization of a*cmah* promoter and provides the first insight into the mechanism(s) regulating the basal promoter activity of *pcmah*. Identification and analysis of *cis-*acting elements and interacting transcription factors, which are important for *pcmah* promoter activity, are essential for understanding the mechanisms of *pcmah* transcription regulation.

Tissue-specific gene expression characterized by transcripts incorporating alternative 5'ends has been shown to be under the control of multiple promoters (Nakamuta, M., et al. 1995, Rajagopalan, S. et al. 1993, Tolner, B. et al. 1998). Our previous study demonstrated that *pcmah* has two 5'alternatively spliced forms, 5'*pcmah*-1 and -2, and has a tissue-specific expression pattern, with 5'*pcmah*-1 mainly expressed in the intestine and5'*pcmah*-2 expressed in most tissues (Song, K.H., et al. 2010). These tissue-specific expression patterns of the two alternative mRNA forms is consistent with the mRNA expression patterns in PK15 cells originating from the pig kidney and IPI-2I cells originating from the pig small intestine (Fig. 1B). Here, we isolated two promoter regions of P1 and P2, which are upstream regions of the 5'*pcmah*-1 and 5'*pcmah*-2 forms, respectively. Luciferase reporter assay results demonstrated that the P2 promoter was highly active in both PK15 and IPI-2I cells, while the P1 promoter was mainly active in IPI-2I cells (Fig. 2). These results correspond with the mRNA expression patterns of the two transcriptsshown in Fig. 1B, suggesting that these alternative promoters might confer differences in regulation of *pcmah* transcription in pig tissues, such as the kidney and small intestine. On the other hand, PK15 cells displayed basal activity of the P1 promoter, even though *pcmah*-1 mRNA expression by promoter P1 was scarcely detected. Here, PK15 cells were used to further investigate the elements related to basal promoter activity. 5' deletion analysis of the P2 promoter suggested that the region between 432 and 334 contains negative regulatory elements, whereas the region between 334 and 210 haspositive regulatory elements (Fig. 4). In addition, 5' deletion analysis of the P1 promoter region suggested that three regions (1100–700, 542–260 and 260–223) act as positive regulatory regions, and that the region between 260–223 isthe basal P1 promoter for 5'*pcmah*-1 expression (Fig. 4).

As shown in Fig. 3, both *pcmah* promoters lack the typical TATA boxes. Instead, two GC boxes were observed in the P2 promoter, but not in the P1 promoter. The TATA-less promoter structure is found in most common housekeeping genes including glycan synthesis- and hydroxylase-related genes (Kim, S.W. et al. 2003, Kim, S.W., et al. 2002, Lundell, K. 2004, Okuda, T. 2008, Smale, S.T. 1997). Moreover, Sp1 is necessary for the transcription of TATA-less genes (Pugh, B.F. 1990). Sp1 ubiquitously expressed in mammalian cells, specifically bind to GC boxes, and is cis-acting DNA regulatory elements required for the transcriptional regulation of many housekeeping, tissue-specific and inducible genes (Li, L., et al. 2004). From analysis through the transcription factor binding prediction algorithm, the *pcmah* P1 and P2 promoter regions were found to contain many GC boxes or Sp1 binding sequences controlled by the Sp1 transcription factor family in their proximal promoters (Fig. 3). The P2 promoter region contains three independent Sp1 binding sites (P2-Sp1a, P2-Sp1b and P2-Sp1c), while the P1 promoter region contains two overlapping Sp1 binding sites (P1-Sp1a and P1-Sp1b). In this study, we focused on*pcmah* regulation by Sp1 among the many transcription factors involved in the regulation of *pcmah* promoter activity. Mutation of the Sp1 binding sites resulted in the reduction of luciferase activity in P2-334 and P1-542, indicating that in the proximal promoter region, Sp1 binding sites are important inregulating the basal level of *pcmah* expression (Fig. 5). In addition, to confirm whether the Sp1 family participates in the transcriptional regulation of *pcmah*, PK15 cells were treated with mithramycin A, a well-known GC-specific DNA binding drug, which prevents the binding of Sp1 family transcription factors totheir specific sites (Blume, S.W. et al. 1991). As expected, 25 nM mithramycin A was sufficient to reduce the promoter activity in P2-334 of the P2 promoter to less than 26%, whereas 50 nM mithramycin A reduced the promoter activity in P1-542 of the P1 promoter to less than 22% (Fig. 6). These results support the suggestion that the different sensitivies for mithramycin A between P1-542 of the P1 promoter and P2-334 of the P2 promoter can be attributed to the number of Sp1 binding sites located on these promoter regions. With respect to the Sp1 binding capacity of both promoters, EMSA analysis revealed that probes containing each Sp1 binding site successfully bound to components of PK15 nuclear extracts (Figs. 7A and C). Additionally, from the EMSA super-shift assay using anti-Sp1 antibody, Sp1 bound to the DNA-nuclear protein complex for all probes containing anSp1 binding site, as Sp1 binding has clearly disappeared (Figs. 7B and D), indicating that the transcription factor Sp1 is a key component of the binding activity derived from PK15 nuclear extracts. These results indicate that Sp1 binds to its putative binding sites on the P1 and P2 promoter regions of *pcmah* to positively regulate promoter activity. Although it is clear that Sp1 is a key regulator in *pcmah* transcription, other transcription factor(s) could also be involved in *pcmah* transcriptional regulation. Although it is clear that Sp1 is essential in maintaining basal activity of the P1 promoter, more complex regulatory element(s) should be required for full activity. In fact, 5'deletion analysis of the P1 promoter suggests that c-Ets-1 may additionally act as a positive regulatory element, since deletion of regions containing putative Ets-1 binding sites caused a marked decrease in P2 luciferase activity (Fig. 4A). Interestingly, it was reported that Ets transcription factors are widely expressed in developing and mature intestine and are involved in an inflammatory disease (Jedlicka, P. and Gutierrez-Hartmann, A. 2008, Konno, S., Iizuka, M., et al. 2004). Further investigation ofthe role of the Ets-1 binding site on the P1 promoter using intestine-derived cells will be necessary to clarify the tissue-specific or disease-inducible expression of *pcmah*-1 by the P1 promoter.

In conclusion, we isolated two promoter regions, P1 and P2, of the *pcmah* gene. The P2 promoter is active in both PK15 pig kidney cells and IPI-21 pig small intestine cells, and appears to have a higher level of basal activity than the P1 promoter in both cell lines, suggesting that P2 is a housekeeping promoter. In contrast, the P1 promoter is highly active in IPI-2I cells compared to thePK15 cells, where the P1 promoter has only basal activity, suggestive of an intestine-specific promoter. Furthermore, our results demonstrate that the Sp1 transcription factor is akey regulatory element important for the regulation of *pcmah* expression. This study provides us with knowledge of the transcriptional regulation of the CMAH gene and may therefore contribute to better understanding of the role of NeuGc regulation in infection, developmental processes, and tumorigenesis. Further investigatation of the additional roles of Sp1 in the transcriptional regulation of *pcmah* and their correlation with pathological processes is in progress.

Chapter 5
Screening for Xenoantigenic Determinants Formed by Sialyltransferases

Kwon-Ho Song and Cheorl-Ho Kim[*]

Molecular and Cellular Glycobiology Laboratory, Department of Biological Science, SungKyunKwan University, 300 Chunchun-Dong, Jangan-Gu, Suwon City, Kyunggi-Do 440-746, Korea
kwonho@live.co.kr, chkimbio@skku.edu

Abstract. In this study, we established pig kidney (PK15) cell lines which are transfected by human sialyltransferases including human a-2, 3-sialyltransferase (hST3Gal II) and human α-2, 6- sialyltransferase (hST6Gal I, hST6GalNAc IV). The transfected cells were screened by Western blot analysis and Lectin blot analysis. The cytotoxicity and xenoreactivity of the PK 15 cell lines were examined by performing the LDH cytotoxicity assay and FACS analysis. From LDH cytotoxicity assay, cytotoxicity to human serum significantly was increased in hST3Gal II and hST6GalNAc IV transfected PK 15 cells as compared to the control. Moreover, from the FACS analysis, binding capacities of human IgG to ST3Gal II- and ST6GalNAc IV-transfected cells were significantly increased, compared with control cells. However, dramatic changes were not observed in IgM binding assay, in comparison with binding assay of IgG. Therefore, our results suggest that alteration of glycosylation-pattern by sialyltransferases including hST3Gal II and hST6GalNAc IV gene results in increase of human serum-mediated cytotoxicity to pig kidney cells and the increased susceptibilities to cytotoxicity may result from the increased human IgG binding capacity to these cells rather than human IgM.

1 Introduction

Carbohydrateantigens, which exist on glycoproteins and glycosphingolipids of all mammalian cells, play a crucial role in pig-to-human xenotransplantation (Ezzelarab et al., 2005). The galactose-α1,3-galactose (Gal) antigen synthesized by α1,3-galactosyltransferase (α1,3-GT or GGTA1) is expressed on the cell surface of almost all mammals with the exception of humans, apes and Old World monkeys (Galili et al., 1988). The Gal antigen is a major cause of hyper-acute rejection (HAR) in xenotransplantation, mediated by natural antibodies that

---

[*] Corresponding author.

K.-H. Song and C.-H. Kim: *Sialo-Xenoantigenic Glycobiology*, SKKU 1, pp. 43–56.
DOI: 10.1007/978-3-642-34094-9_5   © Springer-Verlag Berlin Heidelberg 2013

directly bind to the Gal antigen (Cooper et al., 1994; Galili, 2001). It was demonstrated that HAR can be avoided by eliminating the Gal antigenthrough producing α1,3-GT knockout pigs (Dai et al., 2002; Lai et al., 2002; Phelps et al., 2003). However, the subsequentrejection phenomenon, called acute vascular rejection (AVR), still remains to be overcome (Bach et al., 1996; Kuwaki et al., 2005). Although the mechanism for AVR is not fully understood, it was suggested that xenoreactive natural antibodies existing in natural human sera play a crucial role in AVR (Ierino et al., 1998; Milland et al., 2006). Among the human natural antibodies, xenoreactive antibodies consist of three immunoglobulin subclasses, IgM, IgG, and IgA (Ezzelarab et al., 2005; Koren et al., 1993). Among these immunoglobulins, anti-Gal IgM, which accounts for about 1-8% of total IgM, is the predominant immunoglobulin involved in HAR, and anti-Gal IgG, which accounts for 1-2.4% of total IgG, plays a central role in AVR (Ezzelarab et al., 2005; McMorrow et al., 1997; Schaapherder et al., 1994).

Sialic acids, which are nine-carbon sugars, are usually found at the non-reducing end of oligosaccharide chains, α2,3- or α2,6-linked to a *β*-D-galactopyranosyl (Gal) residue, or α2,6-linked to a *β*-D-*N*-acetylgalactosaminyl (GalNAc) residue or β-D-*N*-acetylglucosaminyl (GlcNAc) residue of the glycoconjugates (Harduin-Lepers et al., 2001). The biosynthesis of sialylated oligosaccharide chains is catalyzed by the sialyltransferase family, a class of glycosyltransferases which share the same donor substrate, CMP-sialic acid (Dall'Olio and Chiricolo, 2001; Harduin-Lepers et al., 2001). Sialic acid residues generated by various sialyltransferases play a crucial role in cancer and the immune system because sialic acids are involved in cell-cell interactions or cell-cell adhesion molecule recognition (Crocker, 2005; Crocker et al., 2007; Dall'Olio and Chiricolo, 2001). In pig-to-human xenotransplantation, it was shown that among various sialyltransferases, ST3Gal III and ST6Gal I reduce the Gal antigen by competing with α1,3-galactosyltransferase for the common acceptor substrate (Ezzelarab and Cooper, 2005; Koma et al., 2000; Tanemura et al., 1998). During asystematic study on the relationship between sialyltransferase and xenoantigenicity, preliminary observations were made that sialyltransferases, including ST6GalNAc IV, enhance xenoreactivity in *invitro* pig-to-human interaction. However, it was not clearly demonstrated that ST6GalNAc IV induces xenoreactivity. In addition, carbohydrate antigens containing sialic acid such as sialosyl-Tn or Hanganutziu-Deicher (HD) are known as non-Gal antigens against which humans are suggested to have naturally occurring antibodies (Cooper et al., 1994; Ezzelarab, Ayares et al., 2005). Hence, it is necessary to investigate the xenoreactivity caused by sialic acids in pig-to-human xenotransplantation.

In this study, we analyzed the effects of alterations in pig glycosylation patterns caused by various human sialyltransferases on human serum mediated cytotoxicity in pig kidney cells. Our results suggest that ST6GalNAc IV increases xenoantigenicity in pig-to-human xenotransplantation.

2 Materials and Methods

2.1 Cell Culture

Pig kidney cell line (PK15) obtained from the Korean Cell Line Bank (KCLB; Seoul, Korea) was cultured in Dulbecco's modified Eagle's medium (DMEM; WelGENE, Korea) containing 100 units of penicillin-streptomycin per ml and 10% fetal bovine serum (FBS) at 37°C in 5% CO_2 incubator/humidified chamber.

2.2 Construction of Recombinant Expression Vectors

The full-length cDNAs of human sialyltransferases including human β-galactoside α2,3-sialyltransferase 2 and 5 (hST3Gal II and hST3Gal V), human β-galactoside α2,6- sialyltransferase 1 (hST6Gal I), human (α-*N*-acetyl-neuraminyl-2,3-α-galactosyl-α1,3) β-*N*- acetylgalactosaminide-α2,6-sialyltransferase 4 (hST6GalNAc IV) and human α2,8- sialyltransferase (hST8Sia I) were inserted 3' of the CMV promoter into the *Bam*HI/*Xho*I sites of pcDNA^{TM}3.1/myc-His expression vector (Invitrogen, Carlsbad, CA). To clone sialyltransferase genes, PCR was performed using EF-Taq polymerase (SolGent, Korea) with human liver cDNA library as the template and the following primer sets containing restriction enzyme sites (*Bam*HI/*Xho*I) were used: hST6Gal I (NM 173216), 5'-AAGGATCCATGATTCACACCA-3' (sense) and 5'-ACTCTAGAGCAGTGAATGGTC -3' (antisense), hST3Gal II (NM006927), 5'-AAGGATCCATGAAGTGCTCCC-3' (sense) and 5'-ACTCTAGAGTTGCCCCGGTAG-3' (antisense), ST6GalNAc IV (NM 175039) 5'- AAGGATCCATGAAGGCTCCGG-3' (sense) and 5'-ACTCTAGACTCAGTCCTCCAG-3' (antisense). The recombinant expression vectors were confirmed by restriction enzymedigestion and DNA sequencing.

2.3 Establishment of Human Sialyltransferase Transfectants

Transfection of human sialyltransferases into PK15 cells was performed using a WelFect-EX^{TM} PLUS Transfection Reagent (WelGENE, Korea). Briefly, PK15 cells were grown to confluence in DMEM. PK15 cells were grown in serum- free DMEM for 1h before transfection. Purified DNA linearized by *Sal* I (1 *u*g/6-well plates) was added to serum-free DMEM containing 3 *u*g of Enhancer-Q and incubated for 10 min. The DNA/Enhancer-Q mixture was added to serum-free DMEM containing 5 *u*g of WelFect-Ex and incubated for 10 min. The DNA/Enhancer-Q/WelFect-Ex mixture was added to the media where PK15 cells were cultured and incubated for 5 h at 37°C with 5% CO_2. After addition of 20% FBS (final 10% concentration), the transfected cells were incubated overnight at 37°C with 5% CO_2. The following morning, the transfection media was aspirated, and cells were cultured in DMEM containing 10% FBS and 100 units of penicillin-streptomycin per ml for 48 h at 37°C with 5% CO_2. Cells were selected for stable integration of transfected DNA by selection in DMEM containing 0.6

mg/ml of G418 (Gibco) for several days. The establishment of PK15 cell lines transfected by human sialyltransferase genes was confirmed by Western blot analysis.

2.4 Western Blot Analysis

The anti-c-Myc (Santa Cruz) and goat anti-mouse IgG-HRP (Santa Cruz) antibodies were purchased for this study. Cells were harvested by scraping and washed twice with PBS and resuspended in RIPA lysis buffer (20 mM Tris pH 7.5, 150 mM NaCl, 1% Triton X-100, 2 mM EDTA, 10% glycerol, 0.1% SDS, 0.5% deoxycholate) containing protease inhibitor cocktail (1mM Na_3VO_4, 20 ug/ml PMSF, 10 ug/l leupepton, 50 mM NaF). Whole lysates from transfectants were separated by 12% SDS PAGE, transferred onto nitrocellulose membrane and immunoblotted with antibodies. Bands were visualized by ECL.

2.5 Lactate Dehydrogenase (LDH) Release Assay

This assay was performed using the CytoTox96Non-Radioactive Cytotoxicity Assay kit (Promega). For human serum-mediated cytotoxicity, the transfected cells were seeded at a density of 4×10^4 cells in flat bottomed 96-well trays 1 day prior to assay. After 15h incubation, the wells were washed twice with serum-free DMEM to remove LDH, which is present in FBS, and then incubated with NHS diluted with DMEM. The plates were incubated for 2 h at 37°C and the released LDH was then measured at 490 nm by using a VERSA max microplate reader (Molecular Device). The percentage of cytotoxicity was calculated using Equation 1

$$\text{Cytotoxicity} = (E - N - S) / (M \ N \ S) * 100 \tag{1}$$

Where E is the experimentally observed release of LDH activity from the target cells, N is LDH activity in 20% NHS, S is the spontaneous release of LDH activity target cells incubated in the absence of NHS, and M is the maximal release of LDH activity, determined by adding lysis solution (0.9% Triton X-100).

For NK cell mediated cytotoxicity, the target transfected cells were seeded at 4×10^4 cells in flat bottomed 96-well trays 1 day prior to assay. After 15h, serial two-fold diluted NK92 MI cells were added to the target cell plates. After 4h incubation at 37°C, release of LDH was analyzed.

2.6 Immunofluorescence Microscopy Analysis

The transfected cells were seeded at 1×10^5 cells on slide glass in 6-well culture plate, 1 day prior to assay. The cells were fixed with 3.7% formaldehyde and then washed twice with PBS. The cells reacted with biotin-conjugated MAL-II lectin (2 mg/ml, 1:500 diluted by PBS/0.5% BSA, Vector Laboratory) and biotin-conjugated PNA (5 mg/ ml, 1:1000 diluted by PBS/0.5% BSA, Vector

Laboratory). After 30 min, the cells were washed twice with PBS and then reacted with Alexa fluor488-conjugated streptavidin (1:5000 diluted by PBS/0.5% BSA, Invitrogen) for 30 min. The cells were washed in PBS, air-dried, and stained with the DNA-specific fluorochrome 4, 6-diamidino-2-phenylindol (DAPI) for 5 min at 37°C. Finally, the DAPI-stained cells were observed under a fluorescence microscope.

2.7 FACS Analysis

The transfected cells were seeded at 1×10^6 cells in a 10 mm culture dish 1 day prior to assay. The cells were washed twice with PBS and harvested by scraping. For the lectin binding assay, the gathered cells were reacted with biotin-conjugated MAL-II lectin (2 mg/ ml, 1:500 diluted by PBS/0.5% BSA, Vector Laboratory) and biotin-conjugated PNA (5 mg/ ml, 1:1000 diluted by PBS/0.5% BSA, Vector Laboratory). After 30 min, the cells were washed twice with PBS and then reacted with Alexa fluor488-conjugated streptavidin (1:5000 diluted by PBS/0.5% BSA, Invitrogen) for 30 min. For the human IgG or IgM binding assay, the gathered cells were reacted with 100 ul of NHS (1:1 diluted by PBS/0.5% BSA). After 30 min incubation on ice, the cells were washed twice with PBS/0.5% BSA and then incubated with 100 ul of FITC-conjugated goat anti human IgG and IgM (Zymed Laboratories), each diluted 1: 50 with PBS/0.5% BSA, on ice for 30 min. To analyze the amount of Gal antigen, the gathered cells were reacted with 100 ul of FITC-conjugated GSIB4 lectin (Sigma) diluted 1: 200 with PBS/0.5% BSA on ice for 30 min. Stained cells were then washed twice with PBS/0.5% BSA, and analyzed using a FACSCalibur (Becton Dickinson).

3 Results

3.1 Establishment of PK15 Cell Lines Transfected by Human Sialyltransferase Genes

Five genes including hST3Gal II, hST3Gal V (GM3 synthase), hST6Gal I, hST6GalNAc IV and ST8Sia IV (GD3 synthase) were used to investigate the effects of human sialyltransferase genes on human serum-mediated cytotoxicity (Fig. 1). Expression vectors containing the full ORF sequences of human sialyltransferase cDNAs were constructed and transfected into PK15 cell lines. To analyze the stable expression of the sialyltransferases in hST3Gal II-, hST3Gal V-, hST6Gal I-, hST6GalNAc IV-, hST8Sia I-transfected cells, Western blotting was performed using anti-c-Myc antibody (Fig. 2). Sialyltransferase-transfected cells expressed the recombinant protein at high levels compared to untransfected cells or vector-transfected cells.

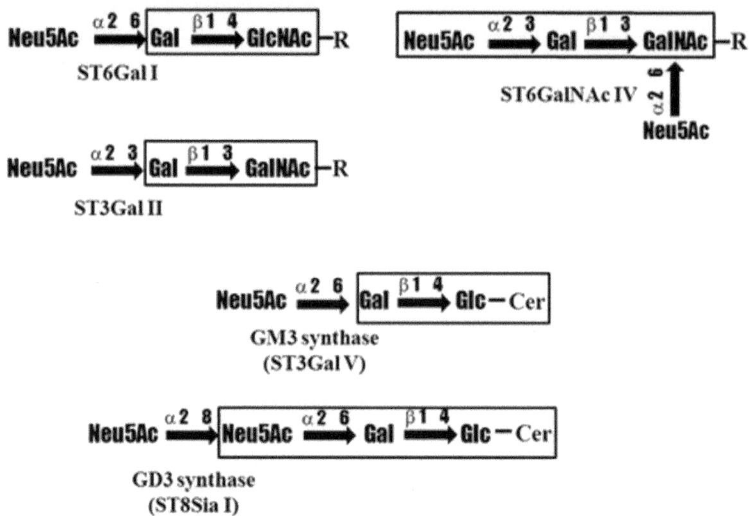

Fig. 1. Sialylation reaction of sialyltransferase used in this study.
Sialylation reaction of each sialyltransferase was shown. Boxes indicate the substrate of each sialyltransferase. R is glycolipid or glycoprotein carrier molecule anchored in the cell membrane.

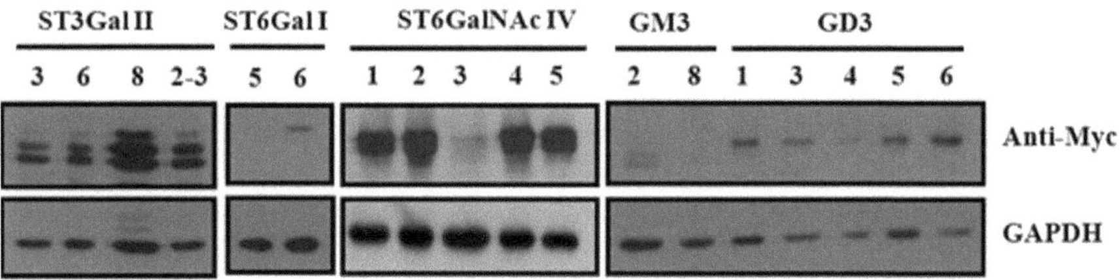

Fig. 2. Establishment of PK15 cell lines transfected by human sialyltransferase genes.
To confirm the establishment of PK15 cell lines transfected by human sialyltransferase genes, Western blot analysis was performed. Whole lysates from transfectants were separated by 12% SDS PAGE and transferred onto nitrocellulose membrane. The anti-c-myc antibody was used as antibodies. As a control for equal protein loading, GAPDH was measure.

3.2 *Human Serum-Mediated Cytotoxicity on Sialyltransferase-Transfected PK15 Cell Lines*

To determine human serum-mediated cytotoxicity, the lactate dehydrogenase (LDH) assay was performed using 20% normal human serum (NHS) which contains natural antibodies and complement. As demonstrated in previous reports (Koma et al., 2000), hST6Gal I-transfected cells slightly decrease in cytotoxicity compared to control vector-transfected cells (Fig. 3). Human serum-mediated cytotoxicity of hST3Gal V- and hST8Sia I-transfected cells did not change compared to the control. In contrast, human serum-mediated cytotoxicity significantly increased in hST3Gal II- and hST6GalNAc IV-transfected cells, to two to three times the level of control cells. These results suggest that an alteration of glycosylation pattern by the hST3Gal II and hST6GalNAc IV gene have an effect on the increase of human serum-mediated cytotoxicity in pig kidney cells. The hST6GalNAc IV-transfected cell line showed the largest increase in human serum mediated cytotoxicity compared to other sialyltransferase gene-transfected cell lines, and thus the hST6GalNAc IV-transfected cells were further analyzed.

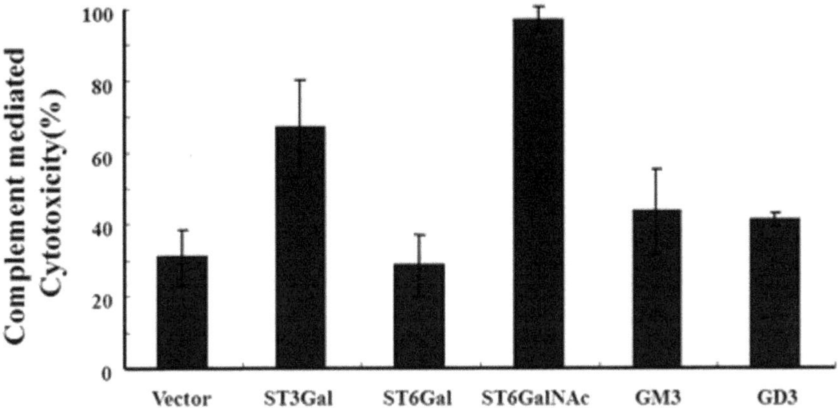

Fig. 3. Human serum-mediated cytotoxicity.
Serum-mediated cytotoxicity ofthe transfectants was estimated by 20% NHS which served as a source of natural antibodies and complement. Percent serum-mediated cytotoxicity was presented. Average values from triplicate assays are shown with S.E values.

3.3   Establishment of hST6GalNAc IV-Transfected ECV304 Cells

Human ST6GalNAc IV has restricted substrate specificity, only utilizing the Neu5Ac2-3Gal1-3GalNAc trisaccharide sequence, which is observed on O-glycosyl proteins (Fig. 4) (Harduin-Lepers, et al., 2001). To further investigate the effects of the hST6GalNAc IV gene on human serum-mediated cytotoxicity, the hST6GalNAc IV gene was transfected into human cell lines (Fig. 5). If the ectopically expressed sialyltransferase is functional in transfected cells, over-expression of the sialyltransferase should cause an alteration of the glycosylation pattern of glycoproteins. To analysis the glycosylation pattern of glycoproteins on the transfected cells, a lectin binding assay was performed using lectins including peanut agglutinin (PNA) and maackia amurensis leukoagglutinin II (MAL II). In a lectin binding assay using PNA, expression of glycoproteins which have a core O-glycan structure was significantly reduced in hST6GalNAc IV-transfected clls (Fig. 6). Binding of terminal disialylated tetrasaccharide Neu5Ac2-3Gal1-3(NeuAc2-6)GalNAc residues-specific MAL-II was increased overall in ST6GalNAc IV-transfected cells compared to the control (Fig. 6). These results indicate that the sialyltransferase-transfected cell lines were successfully established and that the pattern of glycosylation was altered in these cells.

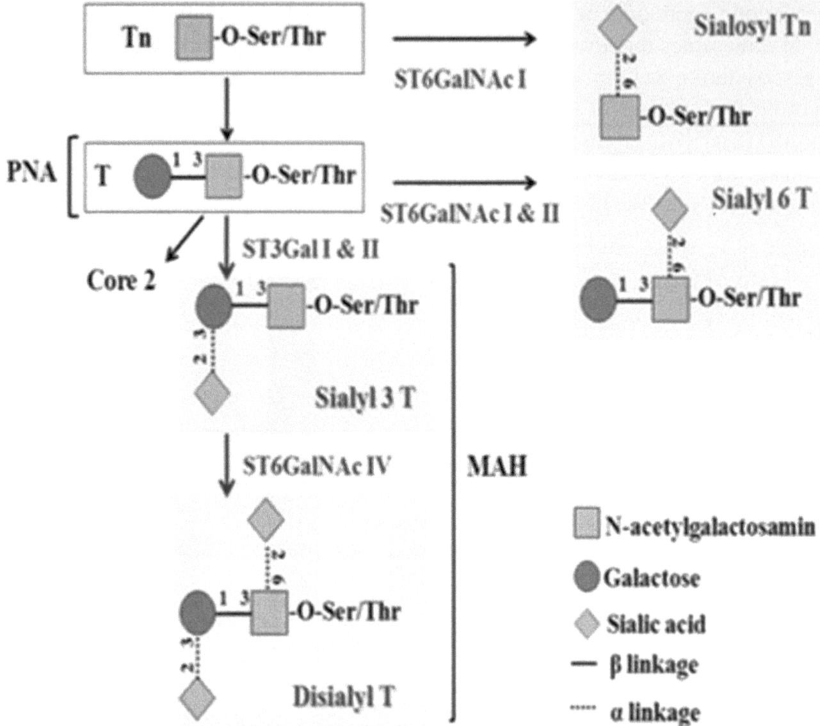

Fig. 4. Sialylation reactions in the initial steps of the O-glycans biosynthesis.
The name of glycoconjugates is indicated beside the glycan structures. The sialyltransferase corresponding to reaction is indicated under or beside the arrow. Lectin PNA or MAH bound to specific glycan structure was represented.

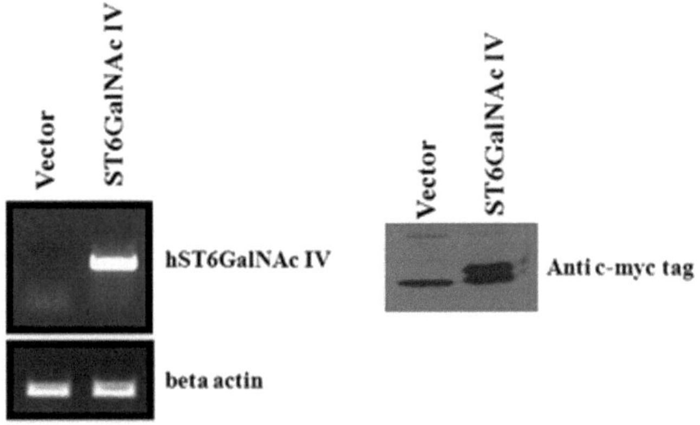

Fig. 5. Establishment of hST6GalNAc IV transfected ECV304 cells.
To confirm the establishment of ECV304 cell lines transfected by hST6GalNAc IV genes, RT-PCR and Western blot analysis was performed.

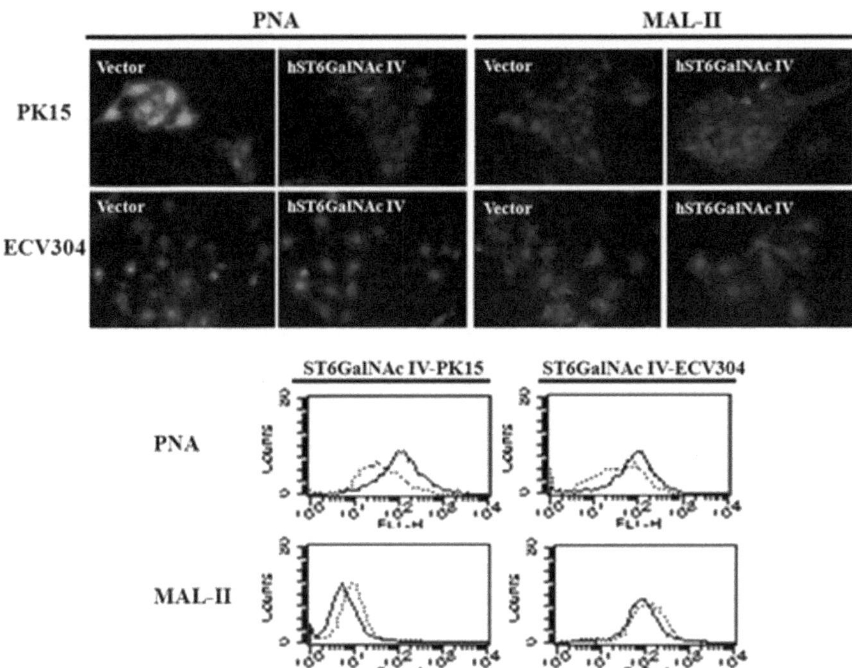

Fig. 6. Lectin binding assay to ST6GalNAc IV transfected cell lines.
To investigate the pattern of glycosylation of hST6GalNAc IV transfected cell lines, a
Lectin binding assay was performed with immunofluorescence microscope (upper) and
FACS (lower). Dotted line indicates the vector transfected cells and thick line indicates the
hST6GalNAc IV transfected cells.

3.4 Human Serum-Mediated Cytotoxicity of hST6GalNAc IV-Transfected PK15 and ECV304 Cell Lines

As shown in Fig. 3, the hST6GalNAc IV-transfected cell line showed the
largest increase in human serum-mediated cytotoxicity (HSMC) than other
sialyltransferase gene-transfected cell lines. We further investigated the HSMC of
hST6GalNAc IV-transfected PK15 cells at various concentrations of NHS, and the
HSMC of hST6GalNAc IV-transfected ECV304 cells in 50% NHS. As shown in
Fig. 7, the HSMC of hST6GalNAc IV-transfected PK15 cells increased in a serum
dependent manner. Also, HSMC was significantly greater in the hST6GalNAc IV-
transfected ECV304 cells compared to the control (Fig. 7).

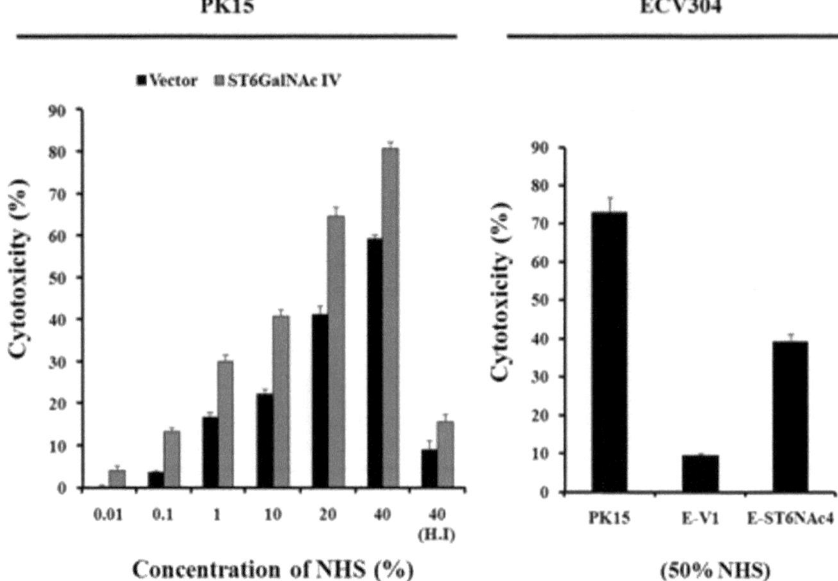

Fig. 7. Human serum-mediated cytotoxicity.
Human serum-mediated cytotoxicity of the transfectants was estimated by NHS which served as a source of natural antibodies and complement. Percent serum-mediated cytotoxicity was presented. Average values from triplicate assays are shown with S.E values.

3.5 Binding of IgG and IgM to hST6GalNAc IV-Transfected PK15 and ECV304 Cell Lines

Xenoreactivity was evaluated by flow cytometry analysis in control and the hST6GalNAc IV-transfected cells. Binding of human IgG to hST6GalNAc IV-transfected cells was significantly increased compared to control cells (Fig. 8). Similar features were also observed in a binding assay for human IgM. However, dramatic changes were not detected when compared with a binding assay of human IgG. Binding of human IgM to hST3Gal II- or hST6GalNAc IV-transfected cells increased slightly compared to the control (Fig. 8). To investigate whether increase of xenoantigenicity was affected by a quantitative change of Gal antigen, we performed a binding analysis of GS-IB4 to hST6GalNAc IV-transfected cells. As expected, binding of GS-IB4 to hST6GalNAc IV -transfected cells was not changed compared to thecontrol (Fig. 9). These results suggest that an alteration of glycosylation pattern by ectopic expression of the hST6GalNAc IV gene affects the binding of IgG rather than IgM in pig kidney cells, and increased xenoreactivity is Gal antigen-independent.

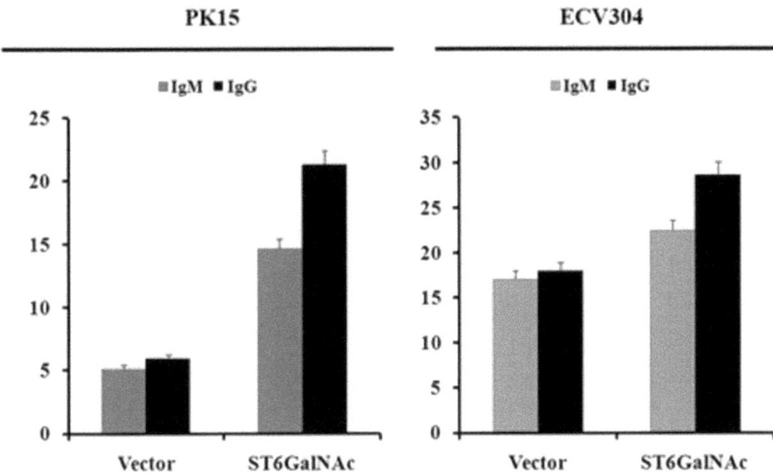

Fig. 8. Binding assay of IgG and IgM to transfected cell lines.
FACS analysis of binding to IgG or IgM in human sera was performed. Cells were reacted with NHS for 30 min and then reacted with FITC-conjugated goat anti human IgG or IgM.

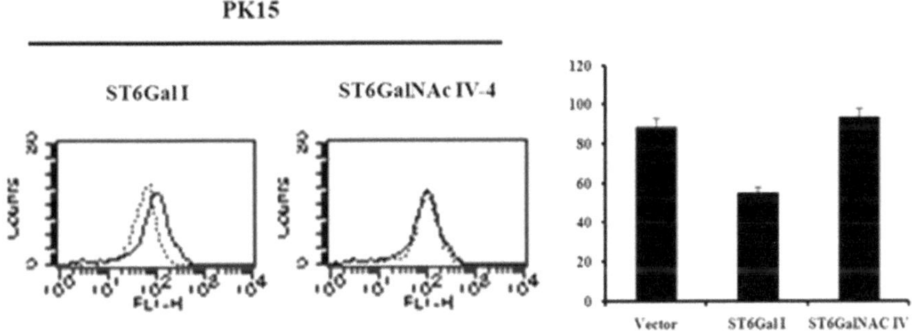

Fig. 9. Binding assay of GS-IB4 lectin to ST6GalNAc IV transfected cell lines.
FITC-conjugated GS-IB4 was used to analyze the amounts of Gal antigen. Stained cells were analyzed by FACS. Right graphs represent results of FACS as the mean fluorescence intensity (MFI) and S.E.M. three independent experiments.

3.6 NK Cell-Mediated Cytotoxicity

To test the role of alteration of glycosylation pattern by the hST6GalNAc IV gene in NK cell cytotoxicity, hST6GalNAc IV-transfected PK15 and ECV304 cells were used as targets for NK-92 MI cells. NK-mediated cytotoxicity was significantly increased in hST6GalNAc IV-transfected PK15 cells (Fig. 10). In general, the PK15 cells are xenogenic for human NK-92 MI cells, but human ECV304 cells are not. However, when effecter NK cells andtarget cells were reacted at a 5:1 ratio, NK cell-mediated cytotoxicity was elevated in hST6GalNAC IV-transfected ECV304 cells compared to the control (Fig. 10). These results indicate that carbohydrate antigens newly synthesized by ST6GalNAc IV-transfected cells are directly recognized.

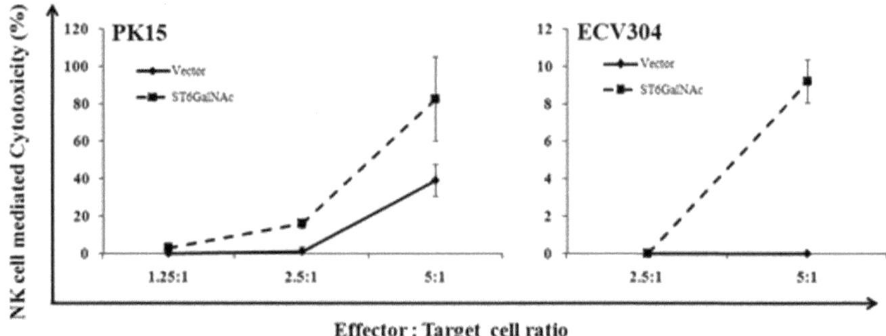

Fig. 10. NK cell mediated cytotoxicity.
NK cell mediated cytotoxicity was estimated with LDH release assay. Effecter cell is NK-92 MI cells and target cell is PK15 or ECV304 cell lines.

4 Discussion

In pig-to-human xenotransplantation, carbohydrate antigens play a crucial role in antibody-mediated rejection processes such as HAR and AVR (Ezzelarab et al., 2005), because humans have naturally occurring antibodies against these carbohydrate antigens (Cooper, 1998). HAR can be prevented by producing α1,3GT deficient pigs (Lai et al., 2002; Phelps et al., 2003). Even if HAR is prevented, AVR can be induced by very low levels or the absence of anti-Gal antibodies, and develops several days or weeks after transplantation (Schuurman et al., 2003; Yang and Sykes, 2007). Although mechanism of AVR is not unclear, Schuurman et al., and Yang and Sykes`s reports indicated that xenoreactive antibodies against a non-Gal antigen can cause AVR. Humans and possibly non-human primates may have naturally occurring antibodies against the non-Gal antigens including the Thomsen-Friedenreich antigen (Gal1-3GalNAc1-R), Tn antigen (GalNAc1-R), sialosyl Tn antigen (NeuAc2-6GalNAc1-R), and P^K antigen (Gal1-4Gal1-4Blc1-R), among others(Cooper, 1998).

Sialic acids, found at the non-reducing end of oligosaccharide chains of glycoconjugates, play a crucial role in cancer and the immune system (Dall'Olio and Chiricolo, 2001; Crocker et al., 2007; Crocker, 2005). In this study, we have established human sialyltransferase-transfected cell lines, to investigate the effects of various sialyltransferases such as ST6Gal I, ST3Gal II, V, ST6GalNAc IV,orST8Sia IV on human serum mediated cytotoxicity in pig kidney cells. hST6Gal I mediates transfer of the sialic acid residue with a 2,6-linkage to the terminal Gal residue of the Gal1-4GlcNAc disaccharide, and hST3Gal II mediates the transfer of sialic acid residues to a terminal Gal residue of theGal1-3GalNAc oligosaccharide (Harduin-Lepers et al., 2001; Kim et al., 1996). ST6GalNAc IV is known as the sialyltransferase that has the most restricted substrate specificity whichutilizes the Neu5Ac2-3Gal1-3GalNAc tetrasaccharide as a unique substrate (Harduin-Lepers et al., 2000; Harduin-Lepers et al., 2001; Lee et al., 1999). Consistent with previous reports, hST6Gal I-transfected cells did not reduce human serum-mediated cytotoxicity. In contrast, hST3Gal II- and hST6GalNAc IV-transfected cells increase human serum-mediated cytotoxicity. These results indicate that carbohydrate antigens generated by hST3Gal II and ST6GalNAc IV may have xenoantigenicity. Especially, the hST6GalNAc IV-transfected cell line was more effective in increasing human serum-mediated cytotoxicity than any other sialyltransferase gene transfected cell line and thus the hST6GalNAc IV-transfected cells were further analyzed. In general, pig cells are xenogenic to human serum and also human NK cells. Therefore, we investigated the effects of ectopic expression of the hST6GalNAc IV gene on HSMC or NK cell-mediated cytotoxicity in human ECV304 cells. HSMC and NK cell-mediated cytotoxicity were significantly greater in the *pcmah*-transfected ECV304 cells compared to the control, as in the PK15 cells (Figs.7 and 10).

Lectin PNA interacts with galactose 1,3 N-acetylgalactosamin, the core structure of O-glycans (Goldstein and Hayes, 1978) and lectin MAL-II interacts with the terminal disialylated tetrasaccharide Neu5Acα2-3Gal1-3(NeuAcα2-6)GalNAc of O-glycans. To investigate whether sialyltransferases successfully function in sialyl-transfected cells, a lectin binding analysis was performed. A lectin binding assay of ST6GalNAc IV transfectants revealed that glycosylation patterns of whole proteins were altered in comparison to the control vector-transfected cells. Binding of core structure-specific PNA was reduced in all transfectants compared to the control in ST6GalNAc IV-transfected cells (Fig. 6). In contrast, binding of the terminal disialylated tetrasaccharide Neu5Acα2-3Gal1-3(NeuAcα2-6)GalNAc residue-specific MAL-II was increased as a whole in ST6GalNAc IV-transfected cells when compared to the control (Fig. 6).

Previous data suggested that anti-Gal IgM is the predominant immunoglobulin involved in HAR and thatanti-Gal IgG plays a central role in AVR (Ezzelarab et al., 2005; McMorrow et al., 1997; Schaapherder et al., 1994). Therefore, a human IgM, IgG binding assay to hST6GalNAc IV- transfected cells was measured by flowcytometry (Fig. 8). Human IgG binding to these cells was significantly increased in hST6GalNAc IV-transfected cells compared to control cells (Fig. 8). Similar results were observed for the IgM binding assay. However, dramatic changes were not detected for the IgG binding assay (Fig. 8). It is formally possible that the changesinxenoantigenicity were affected by a quantitative change inα-Gal epitope, sinceit was demonstrated that ST3Gal III and ST6Gal I decrease xenoantigenicity by reducing the Gal antigen. However, overexpression of the hST6GalNAc IV gene did not have an effect on expression of the Gal antigen. These results suggest that an alteration of pig glycosylation pattern by hST6GalNAc IV has effects on human IgG binding rather than human IgM binding in pig kidney cells and is Gal antigen-independent.

In this study, we investigated the effects of an alteration of pig glycosylation pattern, caused by ectopic expression of human sialyltransferase including hST6GalNAc IV on human serum mediated cytotoxicity in pig kidney cells. Our results suggest that ST6GalNAc IV-transfected cells are susceptible to human serum and human NK cells, resulting in cytotoxicity, and this increased susceptibility may result from increased capacity to bind to human IgG rather than human IgM or direct recognition ofNK cells. ST6GalNAc IV was suggested to be closely related toglycansynthesis of O-glycoproteins (Giordanengo et al., 1997; Harduin-Lepers et al., 2000 Harduin-Lepers et al., 2001; Kim et al., 1996; Kitagawa and Paulson, 1994; Lee et al., 1999). Therefore, carbohydrate antigens newly synthesized by hST6GalNAc IV-transfected cells are potentially believed to be new xenoreactive elements, although it is not at present certain. We are currently investigating what xenoreactive antigens are generated by hST6GalNAc IV in pig kidney cells.

References

Angata, T., Varki, A.: Chemical diversity in the sialic acids and related alpha-keto acids: an evolutionary perspective. Chem. Rev. 102, 439–469 (2002)

Bach, F.H., Turman, M.A., Vercellotti, G.M., Platt, J.L., Dalmasso, A.P.: Accommodation: a working paradigm for progressing toward clinical discordant xenografting. Transplant. Proc. 23, 205–207 (1991)

Bach, F.H., Winkler, H., Ferran, C., Hancock, W.W., Robson, S.C.: Delayed xenograft rejection. Immunol. Today 17, 379–384 (1996)

Blume, S.W., Snyder, R.C., Ray, R., Thomas, S., Koller, C.A., Miller, D.M.: Mithramycin inhibits SP1 binding and selectively inhibits transcriptional activity of the dihydrofolate reductase gene in vitro and in vivo. J. Clin. Invest. 88, 1613–1621 (1991)

Buscher, H.P., Casals-Stenzel, J., Schauer, R., Mestres-Ventura, P.: Biosynthesis of N-glycolylneuraminic acid in porcine submandibular glands. Subcellular site of hydroxylation of N-acetylneuraminic acid in the course of glycoprotein biosynthesis. Eur. J. Biochem. 77, 297–310 (1977)

Byres, E., Paton, A.W., Paton, J.C., Löfling, J.C., Smith, D.F., Wilce, M.C., Talbot, U.M., Chong, D.C., Yu, H., Huang, S., et al.: Incorporation of a non-human glycan mediates human susceptibility to a bacterial toxin. Nature 456, 648–652 (2008)

Chou, H.H., Takematsu, H., Diaz, S., Iber, J., Nickerson, E., Wright, K.L., Muchmore, E.A., Nelson, D.L., Warren, S.T., Varki, A.: A mutation in human CMP-sialic acid hydroxylase occurred after the Homo-Pan divergence. Proc. Natl. Acad. Sci. U S A 95, 11751–11756 (1998)

Chung, T.W., Lee, Y.C., Ko, J.H., Kim, C.H.: Hepatitis B Virus X protein modulates the expression of PTEN by inhibiting the function of p53, a transcriptional activator in liver cells. Cancer Res. 63, 3453–3458 (2003)

Cooper, D.K.: Xenoantigens and xenoantibodies. Xenotransplantation 5, 6–17 (1998)

Cooper, D.K., Koren, E., Oriol, R.: Oligosaccharides and discordant xenotransplantation. Immunol. Rev. 141, 31–58 (1994)

Crocker, P.R.: Siglecs in innate immunity. Curr. Opin. Pharmacol. 5, 431–437 (2005)

Crocker, P.R., Paulson, J.C., Varki, A.: Siglecs and their roles in the immune system. Nat. Rev. Immunol. 7, 255–266 (2007)

Dai, Y., Vaught, T.D., Boone, J., Chen, S.H., Phelps, C.J., Ball, S., Monahan, J.A., Jobst, P.M., McCreath, K.J., Lamborn, A.E., et al.: Targeted disruption of the alpha1,3-galactosyltransferase gene in cloned pigs. Nat. Biotechnol. 20, 251–255 (2002)

Dall'Olio, F., Chiricolo, M.: Sialyltransferases in cancer. Glycoconj. J. 18, 841–850 (2001)

Deicher, H.: Über die Erzeugung heterospezifischer Hämagglutinine durch Injektion artfremden Serums. Z. Hyg. 106, 561–579 (1926)

Devine, P.L., Clark, B.A., Birrell, G.W., Layton, G.T., Ward, B.G., Alewood, P.F., McKenzie, I.F.: The breast tumor-associated epitope defined by monoclonal antibody 3E1.2 is an O-linked mucin carbohydrate containing N-glycolylneuraminic acid. Cancer Res. 51, 5826–5836 (1991)

Ezzelarab, M., Ayares, D., Cooper, D.K.: Carbohydrates in xenotransplantation. Immunol. Cell. Biol. 83, 396–404 (2005)

Ezzelarab, M., Cooper, D.K.: Reducing Gal expression on the pig organ - a retrospective review. Xenotransplantation 12, 278–285 (2005)

Galili, U.: The alpha-gal epitope (Gal alpha 1-3Gal beta 1-4GlcNAc-R) in xenotransplantation. Biochimie. 83, 557–563 (2001)

Galili, U., Shohet, S.B., Kobrin, E., Stults, C.L., Macher, B.A.: Man, apes, and Old World monkeys differ from other mammals in the expression of alpha-galactosyl epitopes on nucleated cells. J. Biol. Chem. 263, 17755–17762 (1988)

Galili, U.: Significance of anti-Gal IgG in chronic xenograft rejection. Transplant. Proc. 31, 940–941 (1999)

Giordanengo, V., Bannwarth, S., Laffont, C., Van Miegem, V., Harduin-Lepers, A., Delannoy, P., Lefebvre, J.C.: Cloning and expression of cDNA for a human Gal(beta1-3)GalNAc alpha2,3-sialyltransferase from the CEM T-cell line. Eur. J. Biochem. 247, 558–566 (1997)

Goldstein, I.J., Hayes, C.E.: The lectins: carbohydrate-binding proteins of plants and animals. Adv. Carbohydr. Chem. Biochem. 35, 127–340 (1978)

Hanganutziu, M.: Hémagglutinines hétérogénétiques après injection de sérum de cheval. C.R. Séances Soc. Biol. 91, 1457–1459 (1924)

Harduin-Lepers, A., Stokes, D.C., Steelant, W.F., Samyn-Petit, B., Krzewinski-Recchi, M.A., Vallejo-Ruiz, V., Zanetta, J.P., Auge, C., Delannoy, P.: Cloning, expression and gene organization of a human Neu5Ac alpha 2-3Gal beta 1-3GalNAc alpha 2,6-sialyltransferase: hST6GalNAcIV. Biochem. J. 352(Pt.1), 37–48 (2000)

Harduin-Lepers, A., Vallejo-Ruiz, V., Krzewinski-Recchi, M.A., Samyn-Petit, B., Julien, S., Delannoy, P.: The human sialyltransferase family. Biochimie 83, 727–737 (2001)

Hedlund, M., Padler-Karavani, V., Varki, N.M., Varki, A.: Evidence for a human-specific mechanism for diet and antibody-mediated inflammation in carcinoma progression. Proc. Natl. Acad. Sci. U S A 105, 18936–18941 (2008)

Heissigerova, H., Breton, C., Moravcova, J., Imberty, A.: Molecular modeling of glycosyltransferases involved in the biosynthesis of blood group A, blood group B, Forssman, and iGb3 antigens and their interaction with substrates. Glycobiology 13, 377–386 (2003)

Higashi, H., Naiki, M., Matuo, S., Okouchi, K.: Antigen of "serum sickness" type of heterophile antibodies in human sera: indentification as gangliosides with N-glycolylneuraminic acid. Biochem. Biophys. Res. Commun. 79, 388–395 (1977)

Higashi, H., Hirabayashi, Y., Fukui, Y., Naiki, M., Matsumoto, M., Ueda, S., Kato, S.: Characterization of N-glycolylneuraminic acid-containing gangliosides as tumor-associated Hanganutziu-Deicher antigen in human colon cancer. Cancer Res. 45, 3796–3802 (1985)

Higashi, H., Nishi, Y., Fukui, Y., Ikuta, K., Ueda, S., Kato, S., Fujita, M., Nakano, Y., Taguchi, T., Sakai, S., et al.: Tumor-associated expression of glycosphingolipid Hanganutziu-Deicher antigen in human cancers. Gann. 75, 1025–1029 (1984)

Hirabayashi, Y., Higashi, H., Kato, S., Taniguchi, M., Matsumoto, M.: Occurrence of tumor-associated ganglioside antigens with Hanganutziu-Deicher antigenic activity on human melanomas. Jpn. J. Cancer Res. 78, 614–620 (1987)

Ierino, F.L., Kozlowski, T., Siegel, J.B., Shimizu, A., Colvin, R.B., Banerjee, P.T., Cooper, D.K., Cosimi, A.B., Bach, F.H., Sachs, D.H., et al.: Disseminated intravascular coagulation in association with the delayed rejection of pig-to-baboon renal xenografts. Transplantation 66, 1439–1450 (1998)

Irie, A., Suzuki, A.: CMP-N-Acetylneuraminic acid hydroxylase is exclusively inactive in humans. Biochem. Biophys. Res. Commun. 248, 330–333 (1998)

Irie, A., Koyama, S., Kozutsumi, Y., Kawasaki, T., Suzuki, A.: The molecular basis for the absence of N-glycolylneuraminic acid in humans. J. Biol. Chem. 273, 15866–15871 (1998)

Jedlicka, P., Gutierrez-Hartmann, A.: Ets transcription factors in intestinal morphogenesis, homeostasis and disease. Histol. Histopathol. 23, 1417–1424 (2008)

Karlsson, N.G., Olson, F.J., Jovall, P.A., Andersch, Y., Enerbäck, Hansson, G.C.: Identification of transient glycosylation alterations of sialylated mucin oligosaccharides during infection by the rat intestinal parasite Nippostrongylus brasiliensis. Biochem. J. 350(Pt. 3), 805–814 (2000)

Kawano, T., Koyama, S., Takematsu, H., Kozutsumi, Y., Kawasaki, H., Kawashima, S., Kawasaki, T., Suzuki, A.: Molecular cloning of cytidine monophospho-N-acetylneuraminic acid hydroxylase. Regulation of species- and tissue-specific expression of N-glycolylneuraminic acid. J. Biol. Chem. 270, 16458–16463 (1995)

Kang, Y.J.: Cloning and characterization of pig CMP-N-acetylneuraminic acid hydroxylase gene. Sungkyunkwan university (2007)

Kawashima, I., Ozawa, H., Kotani, M., Suzuki, M., Kawano, T., Gomibuchi, M., Tai, T.: Characterization of ganglioside expression in human melanoma cells: immunological and biochemical analysis. J. Biochem. 114, 186–193 (1993)

Kelm, S., Schauer, R.: Sialic acids in molecular and cellular interactions. Int. Rev. Cytol. 175, 137–240 (1997)

Kim, S.W., Kang, N.Y., Lee, S.H., Kim, K.W., Kim, K.S., Lee, J.H., Kim, C.H., Lee, Y.C.: Genomic structure and promoter analysis of human NeuAc alpha2,3Gal beta1,3GalNAc alpha2,6-sialyltransferase (hST6GalNAc IV) gene. Gene 305, 113–120 (2003)

Kim, S.W., Lee, S.H., Kim, K.S., Kim, C.H., Choo, Y.K., Lee, Y.C.: Isolation and characterization of the promoter region of the human GM3 synthase gene. Biochim. Biophys. Acta. 1578, 84–89 (2002)

Kim, Y.J., Kim, K.S., Kim, S.H., Kim, C.H., Ko, J.H., Choe, I.S., Tsuji, S., Lee, Y.C.: Molecular cloning and expression of human Gal beta 1,3GalNAc alpha 2,3-sialytransferase (hST3Gal II). Biochem. Biophys. Res. Commun. 228, 324–327 (1996)

Kitagawa, H., Paulson, J.C.: Differential expression of five sialyltransferase genes in human tissues. J. Biol. Chem. 269, 17872–17878 (1994)

Koma, M., Miyagawa, S., Koyota, S., Yoshitatsu, M., Miyoshi, S., Matsuda, H., Tsuji, S., Shirakura, R., Taniguchi, N.: Effects of Gal beta 1,4 GlcNAc alpha 2,6-D-Sialyl transferase on swine xenoantigen. Transplant. Proc. 32, 2509–2510 (2000)

Komoda, H., Miyagawa, S., Kubo, T., Kitano, E., Kitamura, H., Omori, T., Ito, T., Matsuda, H., Shirakura, R.: A study of the xenoantigenicity of adult pig islets cells. Xenotransplantation 11, 237–246 (2004)

Konno, S., Iizuka, M., Yukawa, M., Sasaki, K., Sato, A., Horie, Y., Nanjo, H., Fukushima, T., Watanabe, S.: Altered expression of angiogenic factors in the VEGF-Ets-1 cascades in inflammatory bowel disease. J. Gastroenterol. 39, 931–939 (2004)

Koren, E., Neethling, F.A., Richards, S., Koscec, M., Ye, Y., Zuhdi, N., Cooper, D.K.: Binding and specificity of major immunoglobulin classes of preformed human anti-pig heart antibodies. Transpl. Int. 6, 351–353 (1993)

Kuwaki, K., Tseng, Y.L., Dor, F.J., Shimizu, A., Houser, S.L., Sanderson, T.M., Lancos, C.J., Prabharasuth, D.D., Cheng, J., Moran, K., et al.: Heart transplantation in baboons using alpha1,3-galactosyltransferase gene-knockout pigs as donors: initial experience. Nat. Med. 11, 29–31 (2005)

Kozutsumi, Y., Kawano, T., Yamakawa, T., Suzuki, A.: Participation of cytochrome b5 in CMP-N-acetylneuraminic acid hydroxylation in mouse liver cytosol. J. Biochem. 108, 704–706 (1990)

Kyogashima, M., Ginsburg, V., Krivan, H.C.: Escherichia coli K99 binds to N-glycolylsialoparagloboside and N-glycolyl-GM3 found in piglet small intestine. Arch. Biochem. Biophys. 270, 391–397 (1989)

Lai, L., Kolber-Simonds, D., Park, K.W., Cheong, H.T., Greenstein, J.L., Im, G.S., Samuel, M., Bonk, A., Rieke, A., Day, B.N., et al.: Production of alpha-1,3-galactosyltransferase knockout pigs by nuclear transfer cloning. Science 295, 1089–1092 (2002)

Lee, Y.C., Kaufmann, M., Kitazume-Kawaguchi, S., Kono, M., Takashima, S., Kurosawa, N., Liu, H., Pircher, H., Tsuji, S.: Molecular cloning and functional expression of two members of mouse NeuAcalpha2,3Galbeta1,3GalNAc GalNAcalpha2,6-sialyltransferase family, ST6GalNAc III and IV. J. Biol. Chem. 274, 11958–11967 (1999)

Li, L., He, S., Sun, J.M., Davie, J.R.: Gene regulation by Sp1 and Sp3. Biochem. Cell. Biol. 82, 460–471 (2004)

Lundell, K.: The porcine taurochenodeoxycholic acid 6alpha-hydroxylase (CYP4A21) gene: evolution by gene duplication and gene conversion. Biochem. J. 378, 1053–1058 (2004)

Malykh, Y.N., Shaw, L., Schauer, R.: The role of CMP-N-acetylneuraminic acid hydroxylase in determining the level of N-glycolylneuraminic acid in porcine tissues. Glycoconj. J. 15, 885–893 (1998)

Malykh, Y.N., Schauer, R., Shaw, L.: N-Glycolylneuraminic acid in human tumours. Biochimie 83, 623–634 (2001)

Malykh, Y.N., King, T.P., Logan, E., Kelly, D., Schauer, R., Shaw, L.: Regulation of N-glycolylneuraminic acid biosynthesis in developing pig small intestine. Biochem. J. 370, 601–607 (2003)

Martensen, I., Schauer, R., Shaw, L.: Cloning and expression of a membrane-bound CMP-N-acetylneuraminic acid hydroxylase from the starfish Asterias rubens. Eur. J. Biochem. 268, 5157–5166 (2001)

Matsunami, K., Miyagawa, S., Nakagawa, K., Otsuka, H., Shirakura, R.: Molecular cloning of pigGnT-I and I.2: an application to xenotransplantation. Biochem. Biophys. Res. Commun. 343, 677–683 (2006)

McMorrow, I.M., Comrack, C.A., Sachs, D.H., DerSimonian, H.: Heterogeneity of human anti-pig natural antibodies cross-reactive with the Gal(alpha1,3)Galactose epitope. Transplantation 64, 501–510 (1997)

Merrick, J.M., Zadarlik, K., Milgrom, F.: Characterization of the Hanganutziu-Deicher (serum-sickness) antigen as gangliosides containing n-glycolylneuraminic acid. Int. Arch. Allergy. Appl. Immunol. 57, 477–480 (1978)

Milland, J., Christiansen, D., Sandrin, M.S.: Alpha1,3-galactosyltransferase knockout pigs are available for xenotransplantation: are glycosyltransferases still relevant? Immunol. Cell. Biol. 83, 687–693 (2005)

Milland, J., Christiansen, D., Lazarus, B.D., Taylor, S.G., Xing, P.X., Sandrin, M.S.: The molecular basis for galalpha(1,3)gal expression in animals with a deletion of the alpha1,3galactosyltransferase gene. J. Immunol. 176, 2448–2454 (2006)

Miwa, Y., Kobayashi, T., Nagasaka, T., Liu, D., Yu, M., Yokoyama, I., Suzuki, A., Nakao, A.: Are N-glycolylneuraminic acid (Hanganutziu-Deicher) antigens important in pig-to-human xenotransplantation? Xenotransplantation 11, 247–253 (2004)

Morozumi, K., Kobayashi, T., Usami, T., Oikawa, T., Ohtsuka, Y., Kato, M., Takeuchi, O., Koyama, K., Matsuda, H., Yokoyama, I., Takagi, H.: Significance of histochemical expression of Hanganutziu-Deicher antigens in pig, baboon and human tissues. Transplant. Proc. 31, 942–944 (1999)

Muchmore, E.A., Milewski, M., Varki, A., Diaz, S.: Biosynthesis of N-glycolyneuraminic acid. The primary site of hydroxylation of N-acetylneuraminic acid is the cytosolic sugar nucleotide pool. J. Biol. Chem. 264, 20216–20223 (1989)

Muchmore, E.A.: Developmental sialic acid modifications in rat organs. Glycobiology 2, 337–343 (1992)

Naito, Y., Takematsu, H., Koyama, S., Miyake, S., Yamamoto, H., Fujinawa, R., Sugai, M., Okuno, Y., Tsujimoto, G., Yamaji, T., et al.: Germinal center marker GL7 probes activation-dependent repression of N-glycolylneuraminic acid, a sialic acid species involved in the negative modulation of B-cell activation. Mol. Cell Biol. 27, 3008–3022 (2007)

Nakamuta, M., Oka, K., Krushkal, J., Kobayashi, K., Yamamoto, M., Li, W.H., Chan, L.: Alternative mRNA splicing and differential promoter utilization determine tissue-specific expression of the apolipoprotein B mRNA-editing protein (Apobec1) gene in mice. Structure and evolution of Apobec1 and related nucleoside/nucleotide deaminases. J. Biol. Chem. 270, 13042–13056 (1995)

Nguyen, D.H., Tangvoranuntakul, P., Varki, A.: Effects of natural human antibodies against a nonhuman sialic acid that metabolically incorporates into activated and malignant immune cells. J. Immunol. 175, 228–236 (2005)

Nystedt, J., Anderson, H., Hirvonen, T., Impola, U., Jaatinen, T., Heiskanen, A., Blomqvist, M., Satomaa, T., Natunen, J., Saarinen, J., et al.: Human CMP-N-acetylneuraminic acid hydroxylase is a novel stem cell marker linked to stem cell-specific mechanisms. Stem Cells 28, 258–267 (2010)

Okuda, T., Nakayama, K.: Identification and characterization of the human Gb3/CD77 synthase gene promoter. Glycobiology 18, 1028–1035 (2008)

Omori, T., Nishida, T., Komoda, H., Fumimoto, Y., Ito, T., Sawa, Y., Gao, C., Nakatsu, S., Shirakura, R., Miyagawa, S.: A study of the xenoantigenicity of neonatal porcine islet-like cell clusters (NPCC) and the efficiency of adenovirus-mediated DAF (CD55) expression. Xenotransplantation 13, 455–464 (2006)

Oriol, R., Ye, Y., Koren, E., Cooper, D.K.C.: Carbohydrate antigens of pig tissues reacting with human natural antibodies as potential targets for hyperacute vascular rejection in pig-to-man organ xenotransplantation. Transplantation 56, 1433–1442 (1993)

Phelps, C.J., Koike, C., Vaught, T.D., Boone, J., Wells, K.D., Chen, S.H., Ball, S., Specht, S.M., Polejaeva, I.A., Monahan, J.A., et al.: Production of alpha 1,3-galactosyltransferase-deficient pigs. Science 299, 411–414 (2003)

Pörtner, A., Peter-Katalinić, J., Brade, H., Unland, F., Büntemeyer, H., Müthing, J.: Structural characterization of gangliosides from resting and endotoxin-stimulated murine B lymphocytes. Biochemistry 32, 12685–12693 (1993)

Pugh, B.F., Tjian, R.: Mechanism of transcriptional activation by Sp1: evidence for coactivators. Cell 61, 1187–1197 (1990)

Rajagopalan, S., Wan, D.F., Habib, G.M., Sepulveda, A.R., McLeod, M.R., Lebovitz, R.M., Lieberman, M.W.: Six mRNAs with different 5' ends are encoded by a single gamma-glutamyltransferase gene in mouse. Proc. Natl. Acad. Sci. U S A 90, 6179–6183 (1993)

Sandrin, M.S., Fodor, W.L., Mouhtouris, E., Osman, N., Cohney, S., Rollins, S.A., Guilmette, E.R., Setter, E., Squinto, S.P., McKenzie, I.F.: Enzymatic remodelling of the carbohydrate surface of a xenogenic cell substantially reduces human antibody binding and complement-mediated cytolysis. Nat. Med. 1, 1261–1267 (1995)

Schauer, R., Wember, M.: Hydroxylation and O-acetylation of N-acetylneuraminic acid bound to glycoproteins of isolated subcellular membranes from porcine and bovine submaxillary glands. Hoppe Seylers Z Physiol. Chem. 352, 1282–1290 (1971)

Shaw, L., Schauer, R.: Detection of CMP-N-acetylneuraminic acid hydroxylase activity in fractionated mouse liver. Biochem. J. 263, 355–363 (1989)

Schauer, R., de Freese, A., Gollub, M., Iwersen, M., Kelm, S., Reuter, G., Schlenzka, W., Vandamme-Feldhaus, V., Shaw, L.: Functional and biosynthetic aspects of sialic acid diversity. Indian J. Biochem. Biophys. 34, 131–141 (1997)

Schaapherder, A.F., Daha, M.R., te Bulte, M.T., van der Woude, F.J., Gooszen, H.G.: Antibody-dependent cell-mediated cytotoxicity against porcine endothelium induced by a majority of human sera. Transplantation 57, 1376–1382 (1994)

Schlenzka, W., Shaw, L., Kelm, S., Schmidt, C.L., Bill, E., Trautwein, A.X., Lottspeich, F., Schauer, R.: CMP-N-acetylneuraminic acid hydroxylase: the first cytosolic Rieske iron-sulphur protein to be describedin Eukarya. FEBS Lett. 385, 197–200 (1996)

Schlenzka, W., Shaw, L., Schneckenburger, P., Schauer, R.: Purification and characterization of CMP-N-acetylneuraminic acid hydroxylase from pig submandibular glands. Glycobiology 4, 675–683 (1994)

Schuurman, H.J., Cheng, J., Lam, T.: Pathology of xenograft rejection: a commentary. Xenotransplantation 10, 293–299 (2003)

Shaw, L., Schauer, R.: The biosynthesis of N-glycoloylneuraminic acid occurs by hydroxylation of the CMP-glycoside of N-acetylneuraminic acid. Biol. Chem. Hoppe Seyler 369, 477–486 (1988)

Shibuya, N., Goldstein, I.J., Broekaert, W.F., Nsimba-Lubaki, M., Peeters, B., Peumans, W.J.: The elderberry (Sambucus nigra L.) bark lectin recognizes the Neu5Ac(alpha 2-6)Gal/GalNAc sequence. J. Biol. Chem. 262, 1596–1601 (1987)

Smale, S.T.: Transcription initiation from TATA-less promoters within eukaryotic protein-coding genes. Biochim. Biophys. Acta 1351, 73–88 (1997)

Song, K.H., Kang, Y.J., Jin, U.H., Park, Y.I., Kim, S.M., Seong, H.H., Hwang, S., Yang, B.S., Im, G.S., Min, K.S., et al.: Cloning and functional characterization of pig CMP-N-acetylneuraminic acid hydroxylase for the synthesis of N-glycolylneuraminic acid as the xenoantigenic determinant in pig-human xenotransplantation. Biochem. J. 427, 179–188 (2010)

Tanemura, M., Miyagawa, S., Ihara, Y., Matsuda, H., Shirakura, R., Taniguchi, N.: Significant downregulation of the major swine xenoantigen by N-acetylglucosaminyltransferase III gene transfection. Biochem. Biophys. Res. Commun. 235, 359–364 (1997)

Tanemura, M., Miyagawa, S., Koyota, S., Koma, M., Matsuda, H., Tsuji, S., Shirakura, R., Taniguchi, N.: Reduction of the major swine xenoantigen, the alpha-galactosyl epitope by transfection of the alpha2,3-sialyltransferase gene. J. Biol. Chem. 273, 16421–16425 (1998)

Tangvoranuntakul, P., Gagneux, P., Diaz, S., Bardor, M., Varki, N., Varki, A., Muchmore, E.: Human uptake and incorporation of an immunogenic nonhuman dietary sialic acid. Proc. Natl. Acad. Sci. U S A 100, 12045–12050 (2003)

Taylor, S.G., McKenzie, I.F., Sandrin, M.S.: Characterization of the rat alpha(1,3)galactosyltransferase: evidence for two independent genes encoding glycosyltransferases that synthesize Galalpha(1,3)Gal by two separate glycosylation pathways. Glycobiology 13, 327–337 (2003)

Tolner, B., Roy, K., Sirotnak, F.: Structural analysis of the human RFC-1 gene encoding a folate transporter reveals multiple promoters and alternatively spliced transcripts with 5' end heterogeneity. Gene 211, 331–341 (1998)

Townsend, R.R., Hardy, M.R., Lee, Y.C.: Separation of oligosaccharides using high-performance anion-exchange chromatography with pulsed amperometric detection. Methods Enzymol. 179, 65–76 (1989)

Traving, C., Schauer, R.: Structure, function and metabolism of sialic acids. Cell Mol. Life. Sci. 54, 1330–1349 (1998)

Wang, W.C., Cummings, R.D.: The immobilized leukoagglutinin from the seeds of Maackia amurensis binds with high affinity to complex-type Asn-linked oligosaccharides containing terminal sialic acid-linked alpha-2,3 to penultimate galactose residues. J. Biol. Chem. 263, 4576–4585 (1988)

Yang, Y.G., Sykes, M.: Xenotransplantation: current status and a perspective on the future. Nat. Rev. Immunol. 7, 519–531 (2007)

Yin, J., Hashimoto, A., Izawa, M., Miyazaki, K., Chen, G.Y., Takematsu, H., Kozutsumi, Y., Suzuki, A., Furuhata, K., Cheng, F.L., et al.: Hypoxic culture induces expression of sialin, a sialic acid transporter, and cancer-associated gangliosides containing non-human sialic acid on human cancer cells. Cancer Res. 66, 2937–2945 (2006)

Varki, A.: Diversity in the sialic acids. Glycobiology 2, 25–40 (1992)

Varki, A.: Biological roles of oligosaccharides: all of the theories are correct. Glycobiology 3, 97–130 (1993)

Varki, A.: Loss of N-glycolylneuraminic acid in humans: Mechanisms, consequences, and implications for hominid evolution. Am. J. Phys. Anthropol. Suppl. 33(Suppl.), 54–69 (2001)

FSC
www.fsc.org
MIX
Papier aus verantwortungsvollen Quellen
Paper from responsible sources
FSC® C105338